Communications
in Computer and Information Science   **705**

Commenced Publication in 2007
Founding and Former Series Editors:
Alfredo Cuzzocrea, Dominik Ślęzak, and Xiaokang Yang

More information about this series at http://www.springer.com/series/7899

Tristan Cazenave · Mark H.M. Winands
Stefan Edelkamp · Stephan Schiffel
Michael Thielscher · Julian Togelius (Eds.)

Computer Games

5th Workshop on Computer Games, CGW 2016
and 5th Workshop on General Intelligence
in Game-Playing Agents, GIGA 2016
Held in Conjunction with the 25th International Conference
on Artificial Intelligence, IJCAI 2016
New York, USA, July 9–10, 2016
Revised Selected Papers

 Springer

Editors
Tristan Cazenave
LAMSADE
Université Paris-Dauphine
Paris Cedex 16
France

Mark H.M. Winands
Department of Data Science
  and Knowledge Engineering
Maastricht University
Maastricht, Limburg
The Netherlands

Stefan Edelkamp
Institute for Artificial Intelligence
Universität Bremen, TAB
Bremen, Bremen
Germany

Stephan Schiffel
Reykjavik University
Reykjavik
Iceland

Michael Thielscher
The University of New South Wales
Sydney, NSW
Australia

Julian Togelius
Department of Computer Science
  and Engineering
New York University
Brooklyn, NY
USA

ISSN 1865-0929   ISSN 1865-0937   (electronic)
Communications in Computer and Information Science
ISBN 978-3-319-57968-9   ISBN 978-3-319-57969-6   (eBook)
DOI 10.1007/978-3-319-57969-6

Library of Congress Control Number: 2017938158

Printed on acid-free paper

This Springer imprint is published by Springer Nature
The registered company is Springer International Publishing AG
The registered company address is: Gewerbestrasse 11, 6330 Cham, Switzerland

Preface

These joint proceedings contain the papers of the Computer Games Workshop (CGW 2016) and the General Intelligence in Game-Playing Agents (GIGA 2016) Workshop, which were both held in New York, USA. These workshops took place on July 9 and 10, 2016, respectively, in conjunction with the 25th International Conference on Artificial Intelligence (IJCAI 2016). These two workshops reflect the large interest in artificial intelligence (AI) research for games.

The Computer and Games Workshop series is an international forum for researchers interested in all aspects of AI and computer game playing. Earlier workshops took place in Montpellier, France (2012), Beijing, China (2013), Prague, Czech Republic (2014), and Buenos Aires, Argentina (2015). For the fifth edition of the Computer Games Workshop, 17 submissions were received in 2016. Each paper was sent to three reviewers. In the end, 11 papers were accepted for presentation at the workshop, of which nine made it into these proceedings. The published papers cover a wide range of topics related to computer games. They collectively discuss five abstract games: Breakthrough, Go, Hex, SameGame, and Werewolf. Additionally, one paper deals with optimization problems such as bus regulations and weak Schur numbers, and two papers are on video games.

The GIGA workshop series has been established to become the major forum for discussing, presenting, and promoting research on general game playing (GGP). It aims at building intelligent software agents that can, given the rules of any game, automatically learn a strategy for playing that game at an expert level without any human intervention. The workshop intends to bring together researchers from subfields of AI to discuss how best to address the challenges and further advance the state of the art of general game-playing systems and generic AI. Following the inaugural GIGA Workshop at IJCAI 2009 in Pasadena (USA), follow-up events took place at IJCAI 2011 in Barcelona (Spain), IJCAI 2013 in Beijing (China), and IJCAI 2015 in Buenos Aires (Argentina). This fifth workshop on General Intelligence in Game-Playing Agents received eight submissions. Each paper was sent to three reviewers. All papers were accepted for presentation at the workshop, but in the end three were accepted for these proceedings. The accepted papers focus on general techniques for automated reasoning about new games and cover the topics of propositional networks, ground instantiations of game rules, and decomposition of game descriptions.

In all, 48% of the submitted papers for both workshops were selected for these proceedings. Here we provide a brief outline of the 12 contributions, in the order in which they appear in the proceedings. They are divided into two parts: the first nine belong to the Computer Games Workshop and the last three to the GIGA Workshop.

Computer Games Workshop

"NeuroHex: A Deep Q-learning Hex Agent," a joint effort by Kenny Young, Gautham Vasan, and Ryan Hayward, considers deep Q-learning for the game of Hex. After supervised initializing, self-play is used to train NeuroHex, an 11-layer convolutional neural network that plays Hex on the 13 × 13 board. Hex is the classic two-player alternate-turn stone placement game played on a rhombus of hexagonal cells in which the winner is the one who connects their two opposing sides. Despite the large action and state space, their system trains a Q-network capable of strong play with no search. After two weeks of Q-learning, NeuroHex achieves win rates of 20.4% as first player and 2.1% as second player against a one-second/move version of MoHex, the current ICGA Olympiad Hex champion. The data suggest further improvement might be possible with more training time.

"Deep or Wide? Learning Policy and Value Neural Networks for Combinatorial Games," by Stefan Edelkamp, raises the question on the availability, the limits, and the possibilities of deep neural networks for other combinatorial games than Go. As a step toward this direction, a value network for Tic-Tac-Toe was trained, providing perfect winning information obtained by retrograde analysis. Next, a policy network was trained for the SameGame, a challenging combinatorial puzzle. Here, the interplay of deep learning with nested rollout policy adaptation (NRPA) is discussed, a randomized algorithm for optimizing the outcome of single-player games. In both cases the observation is that ordinary feed-forward neural networks can perform better than convolutional ones both in accuracy and efficiency.

"Integrating Factorization Ranked Features in MCTS: An Experimental Study" authored by Chenjun Xiao and Martin Müller investigates the problem of integrating feature knowledge learned by the factorization Bradley–Terry model in Monte Carlo tree search (MCTS). The open source Go program Fuego is used as the test platform. Experimental results show that the FBT knowledge is useful in improving the performance of Fuego.

"Nested Rollout Policy Adaptation with Selective Policies," by Tristan Cazenave, discusses nested rollout policy adaptation (NRPA). It is a Monte Carlo search algorithm that has found record-breaking solutions for puzzles and optimization problems. It learns a playout policy online that dynamically adapts the playouts to the problem at hand using more selectivity in the playouts. The idea is applied to three different domains: Bus regulation, SameGame, and weak Schur numbers. For each of these problems, selective policies improve NRPA.

"A Rollout-Based Search Algorithm Unifying MCTS and Alpha-Beta" by Hendrik Baier integrates MCTS and minimax tightly into one rollout-based hybrid search algorithm, MCTS-$\alpha\beta$. The hybrid is able to execute two types of rollouts: MCTS rollouts and alpha-beta rollouts. During the search, all nodes accumulate both MCTS value estimates as well as alpha-beta value bounds. The two types of information are combined in a given tree node whenever alpha-beta completes a deepening iteration rooted in that node by increasing the MCTS value estimates for the best move found by alpha-beta. A single parameter, the probability of executing MCTS rollouts vs. alpha-beta rollouts, makes it possible for the hybrid to subsume both MCTS as well as alpha-beta search as extreme cases, while allowing for a spectrum of new search algorithms in between. Preliminary results in the game of Breakthrough show that the proposed hybrid outperforms its special cases of alpha-beta and MCTS.

"Learning from the Memory of Atari 2600," written by Jakub Sygnowski and Henryk Michalewski, describes the training of neural networks to play the games Bowling, Breakout, and Seaquest using information stored in the memory of a video game console Atari 2600. Four models of neural networks are considered that differ in size and architecture: two networks that use only information contained in the RAM and two mixed networks that use both information in the RAM and information from the screen. In all games the RAM agents are on a par with the benchmark screen-only agent. In the case of Seaquest, the trained RAM-only agents behave even better than the benchmark agent. Mixing screen and RAM does not lead to an improved performance compared with screen-only and RAM-only agents.

"Clustering-Based Online Player Modeling," a joint collaboration by Jason M. Bindewald, Gilbert L. Peterson, and Michael E. Miller, presents a clustering and locally weighted regression method for modeling and imitating individual players. The algorithm first learns a generic player cluster model that is updated online to capture an individual's game-play tendencies. The models can then be used to play the game or for analysis to identify how different players react to separate aspects of game states. The method is demonstrated on a tablet-based trajectory generation game called Space Navigator.

"AI Wolf Contest — Development of Game AI using Collective Intelligence," a joint effort by Fujio Toriumi, Hirotaka Osawa, Michimasa Inaba, Daisuke Katagami, Kosuke Shinoda, and Hitoshi Matsubara, introduces a platform to develop a game-playing AI for a Werewolf competition. First, the paper discusses the essential factors in Werewolf with reference to other studies. Next, a platform for an AI game competition is developed that uses simplified rules to support the development of AIs that can play Werewolf. The paper reports the process and analysis of the results of the competition.

"Semantic Classification of Utterances in a Language-Driven Game," written by Kellen Gillespie, Michael W. Floyd, Matthew Molineaux, Swaroop S. Vattam, and David W. Aha, describes a goal reasoning agent architecture that allows an agent to classify natural language utterances, hypothesize about humans' actions, and recognize their plans and goals. The paper focuses on one module of the architecture, the natural language classifier, and demonstrates its use in a multiplayer tabletop social deception game, One Night Ultimate Werewolf. The results indicate that the system can obtain reasonable performance even when the utterances are unstructured, deceptive, or ambiguous.

GIGA Workshop

"Optimizing Propositional Networks" authored by Chiara F. Sironi and Mark H.M. Winands analyzes the performance of a Propositional Network (PropNet)-based reasoner for interpreting the game rules, written in the Game Description Language (GDL). The paper evaluates four different optimizations for the PropNet structure that can help further increase its reasoning speed in terms of visited game states per second.

"Grounding GDL Game Descriptions" by Stephan Schiffel discusses grounding game descriptions using a state-of-the art answer set programming system as a viable alternative to the GDL specific approach implemented in the GGP-Base framework. The presented system is able to handle more games and is typically faster despite the overhead of transforming GDL into a different format and starting and communicating with a separate process. Furthermore, this grounding of a game description is well-founded theoretically by the transformation into answer set programs. It allows one to optimize the descriptions further without changing their semantics.

"A General Approach of Game Description Decomposition for General Game Playing," a joint effort by Aline Hufschmitt, Jean-Noël Vittaut, and Jean Méhat, presents a general approach for the decomposition of games described in GDL. In the field of general game playing, the exploration of games described in GDL can be significantly sped up by the decomposition of the problem in subproblems analyzed separately. The discussed program can decompose game descriptions with any number of players while addressing the problem of joint moves. This approach is used to identify perfectly separable subgames but can also decompose serial games composed of two subgames and games with compound moves while avoiding reliance on syntactic elements that can be eliminated by simply rewriting the GDL rules. The program has been tested on 40 games, compound or not. It decomposes 32 of them successfully in less than five seconds.

These proceedings would not have been produced without the help of many persons. In particular, we would like to mention the authors and reviewers for their help. Moreover, the organizers of IJCAI 2016 contributed substantially by bringing the researchers together.

January 2017

Tristan Cazenave
Mark H.M. Winands
Stefan Edelkamp
Stephan Schiffel
Michael Thielscher
Julian Togelius

Organization

Program Chairs

Tristan Cazenave	Université Paris-Dauphine, France
Mark H.M. Winands	Maastricht University, The Netherlands
Stefan Edelkamp	University of Bremen, Germany
Stephan Schiffel	Reykjavik University, Iceland
Michael Thielscher	University of New South Wales, Australia
Julian Togelius	New York University, USA

Program Committee

Christopher Archibald	Mississippi State University, USA
Yngvi Björnsson	Reykjavik University, Iceland
Bruno Bouzy	Université Paris-Descartes, France
Tristan Cazenave	Université Paris-Dauphine, France
Stefan Edelkamp	University of Bremen, Germany
Michael Genesereth	Stanford University, USA
Ryan Hayward	University of Alberta, Canada
Hiroyuki Iida	Japan Advanced Institute of Science and Technology, Japan
Nicolas Jouandeau	Université Paris 8, France
Łukasz Kaiser	Université Paris-Diderot, France
Jialin Liu	University of Essex, UK
Richard Lorentz	California State University, USA
Simon Lucas	University of Essex, UK
Jacek Mańdziuk	Warsaw University of Technology, Poland
Jean Méhat	Université Paris 8, France
Martin Müller	University of Alberta, Canada
Diego Perez	University of Essex, UK
Thomas Philip Runarsson	University of Iceland, Iceland
Abdallah Saffidine	University of New South Wales, Australia
Spyridon Samothrakis	University of Essex, UK
Tom Schaul	Google DeepMind, UK
Stephan Schiffel	Reykjavik University, Iceland
Sam Schreiber	Google Inc. USA
Nathan Sturtevant	University of Denver, USA
Olivier Teytaud	Google, Switzerland
Michael Thielscher	University of New South Wales, Australia
Julian Togelius	New York University, USA

Mark H.M. Winands   Maastricht University, The Netherlands
Shi-Jim Yen   National Dong Hwa University, Taiwan

Additional Reviewer

Chiara Sironi

Contents

Computer Games Workshop 2016

NeuroHex: A Deep Q-learning Hex Agent

Kenny Young$^{(\boxtimes)}$, Gautham Vasan, and Ryan Hayward

Department of Computing Science, University of Alberta, Edmonton, Canada
kjyoung@ualberta.ca

Abstract. DeepMind's recent spectacular success in using deep convolutional neural nets and machine learning to build superhuman level agents—e.g. for Atari games via deep Q-learning and for the game of Go via other deep Reinforcement Learning methods—raises many questions, including to what extent these methods will succeed in other domains. In this paper we consider DQL for the game of Hex: after supervised initializing, we use self-play to train NeuroHex, an 11-layer convolutional neural network that plays Hex on the 13×13 board. Hex is the classic two-player alternate-turn stone placement game played on a rhombus of hexagonal cells in which the winner is whomever connects their two opposing sides. Despite the large action and state space, our system trains a Q-network capable of strong play with no search. After two weeks of Q-learning, NeuroHex achieves respective win-rates of 20.4% as first player and 2.1% as second player against a 1-s/move version of MoHex, the current ICGA Olympiad Hex champion. Our data suggests further improvement might be possible with more training time.

1 Motivation, Introduction, Background

1.1 Motivation

DeepMind's recent spectacular success in using deep convolutional neural nets and machine learning to build superhuman level agents—e.g. for Atari games via deep Q-learning and for the game of Go via other deep Reinforcement Learning methods—raises many questions, including to what extent these methods will succeed in other domains. Motivated by this success, we explore whether DQL can work to build a strong network for the game of Hex.

1.2 The Game of Hex

Hex is the classic two-player connection game played on an $n \times n$ rhombus of hexagonal cells. Each player is assigned two opposite sides of the board and a set of colored stones; in alternating turns, each player puts one of their stones on an empty cell; the winner is whomever joins their two sides with a contiguous chain of their stones. Draws are not possible (at most one player can have a winning chain, and if the game ends with the board full, then exactly one player will have such a chain), and for each $n \times n$ board there exists a winning strategy for

© Springer International Publishing AG 2017
T. Cazenave et al. (Eds.): CGW 2016/GIGA 2016, CCIS 705, pp. 3–18, 2017.
DOI: 10.1007/978-3-319-57969-6_1

the 1st player [7]. Solving—finding the win/loss value—arbitrary Hex positions is P-Space complete [11].

Despite its simple rules, Hex has deep tactics and strategy. Hex has served as a test bed for algorithms in artificial intelligence since Shannon and E.F. Moore built a resistance network to play the game [12]. Computers have solved all 9×9 1-move openings and two 10×10 1-move openings, and 11×11 and 13×13 Hex are games of the International Computer Games Association's annual Computer Olympiad [8].

In this paper we consider Hex on the 13×13 board (Fig. 1).

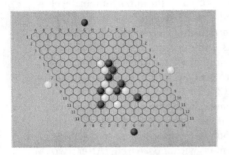

(a) A Hex game in progress. Black wants to join top and bottom, White wants to join left and right.

(b) A finished Hex game. Black wins.

Fig. 1. The game of Hex.

1.3   Related Work

The two works that inspire this paper are [10,13], both from Google DeepMind.

[10] introduces Deep Q-learning with Experience Replay. Q-learning is a reinforcement learning (RL) algorithm that learns a mapping from states to action values by backing up action value estimates from subsequent states to improve those in previous states. In Deep Q-learning the mapping from states to action values is learned by a Deep Neural network. Experience replay extends standard Q-learning by storing agent experiences in a memory buffer and sampling from these experiences every time-step to perform updates. This algorithm achieved superhuman performance on several classic Atari games using only raw visual input.

[13] introduces AlphaGo, a Go playing program that combines Monte Carlo tree search with convolutional neural networks: one guides the search (policy network), another evaluates position quality (value network). Deep reinforcement learning (RL) is used to train both the value and policy networks, which each take a representation of the gamestate as input. The policy network outputs a probability distribution over available moves indicating the likelihood of choosing each move. The value network outputs a single scalar value estimating

$V(S) = P(win|S) - P(loss|S)$, the expected win probability minus the expected loss probability for the current boardstate S. Before applying RL, AlphaGo's network training begins with supervised mentoring: the policy network is trained to replicate moves from a database of human games. Then the policy network is trained by playing full games against past versions of their network, followed by increasing the probability of moves played by the winner and decreasing the probability of moves played by the loser. Finally the value network is trained by playing full games from various positions using the trained policy network, and performing a gradient descent update based on the observed game outcome. Temporal difference (TD) methods—which update value estimates for previous states based on the systems own evaluation of subsequent states, rather than waiting for the true outcome—are not used.

An early example of applying RL with a neural network to games is TD-gammon [15]. There a network trained with TD methods to approximate state values achieved superhuman play. Recent advances in deep learning have opened the doors to apply such methods to more games.

1.4   Overview of this Work

In this work we explore the application of Deep Q-learning with Experience Replay, introduced in [10], to Hex. There are several challenges involved in applying this method, so successful with Atari, to Hex. One challenge is that there are fewer available actions in Atari than in Hex (e.g. there are 169 possible initial moves in 13×13 Hex). Since Q-learning performs a maximization over all available actions, this large number might cause the noise in estimation to overwhelm the useful signal, resulting in catastrophic maximization bias. However in our work we found the use of a convolutional neural network—which by design learns features that generalize over spatial location—achieved good results.

Another challenge is that the reward signal in Hex occurs only at the end of a game, so (with respect to move actions) is infrequent, meaning that most updates are based only on network evaluations without immediate win/loss feedback. The question is whether the learning process will allow this end-of-game reward information to propagate back to the middle and early game. To address this challenge, we use supervised mentoring, training the network first to replicate the action values produced by a heuristic over a database of positions. Such training is faster than RL, and allows the middle and early game updates to be meaningful at the start of Q-learning, without having to rely on end-of-game reward propagating back from the endgame. As with AlphaGo [13], we apply this heuristic only to initialize the network: the reward in our Q-learning is based only on the outcome of the game being played.

The main advantage of using a TD method such as Q-learning over training based only on final game outcomes, as was done with AlphaGo, is data efficiency. Making use of subsequent evaluations by our network allows the system to determine which specific actions are better or worse than expected based on previous training by observing where there is a sudden rise or fall in evaluation. A system that uses only the final outcome can only know that the set of moves made by the winner should be encouraged and those made by the loser discouraged, even though many of the individual moves made by the winner may be bad and many of the individual moves made by the loser may be good. We believe this difference is part of what allows us to obtain promising results using less computing power than AlphaGo.

1.5   Reinforcement Learning

Reinforcement learning is a process that learns from actions that lead to a goal. An agent learns from the environment and makes decisions. Everything that the agent can interact with is called the environment. The agent and environment interact continually: the agent selecting actions and the environment responding to those actions and presenting new situations to the agent. The environment also reports rewards: numerical/scalar values that the agent tries to maximize over time. A complete specification of an environment defines a task, which is one instance of the reinforcement learning problem (Fig. 2).

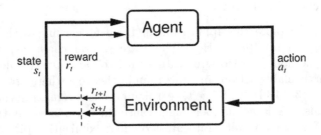

Fig. 2. The agent-environment interaction in reinforcement learning.

The agent and environment interact at each discrete time step (t = 0, 1, 2, 3...). At each time step the agent receives some representation of the environment's state, $s_t \in \mathcal{S}$, where $\mathcal{S}$ is the set of possible states, and on that basis selects an action, $a_t \in \mathcal{A}_t$, where $\mathcal{A}_t$ is the set of actions available in state s_t. One time step later, in part as a consequence of its action, the agent receives a numerical reward, $r_{t+1} \in \mathcal{R}$, and finds itself in a new state s_{t+1}.

The purpose or goal of the agent is formalized in terms of a special reward signal passing from the environment to the agent. At each time step, the reward is a scalar, $r_t \in \mathcal{R}$. Informally, the agent's goal is to maximize the total amount of reward it receives. This means maximizing not immediate reward, but cumulative reward in the long run. The field of reinforcement learning is primarily the study of methods for tackling this challenge.

A RL agent chooses actions according to a policy $\pi(a|s)$ which is a probability distribution over all possible actions for the current state. The policy may be deterministic or stochastic. For a given policy we define the value of a state $v_\pi(S)$ as the expectation value of cumulative reward from state S if we follow π.

$$v_\pi(S) = E_\pi(\sum_{t=1}^{\infty} \gamma^{t-1} r_t | s_0 = S)$$

where γ is a discount factor indicating how much more to credit immediate reward than long term reward; this is generally necessary to ensure reward is finite if the agent-environment interaction continues indefinitely, however it may be omitted if the interaction ends in bounded time, for example in a game of Hex. For a given problem we define the optimal policy π_* (not necessarily unique) as that which produces the highest value in every state. We then denote this highest achievable value as $v_*(S)$. Note that neither $v_\pi(S)$ or $v_*(S)$ are tractable to compute in general, however it is the task of a wide variety of RL algorithms to estimate them from the agent's experience.

Similarly we can define for any policy the action value of each state action pair $q_\pi(S, A)$ which, analogous to $v_\pi(S)$, is defined as the expectation value of cumulative reward from state S if we take action A and follow π after that. Similarly we can define $q_*(S, A)$ as $q_{\pi_*}(S, A)$. Notice that choosing the action with the highest $q_*(S, A)$ in each state is equivalent to following the optimal policy π_*.

See [14] for excellent coverage of these topics and many others pertaining to RL.

1.6   Deep Q-learning

Q-learning is based on the following recursive expression, called a Bellman equation for $q_*(S, A)$.

$$q_*(s_t, a_t) = E(r_{t+1} + \gamma \max_a q_*(s_{t+1}, a) | s_t, a_t)$$

Note that this expression can be derived from the definition of $q_*(s_t, a_t)$. From this expression we formulate an update rule which allows us to iteratively update an estimate of $q_*(S, A)$, typically written $Q_*(S, A)$ or simply $Q(S, A)$ from the agent's stream of experience as follows:

$$Q(s_t, a_t) \xleftarrow{\alpha} r_{t+1} + \gamma \max_a Q(s_{t+1}, a)$$

where in the tabular case (all state action pairs estimated independently) "$\xleftarrow{\alpha}$" would represent moving the left-hand-side value toward the right-hand-side value by some step size α fraction of the total difference, in the function approximation case (for example using a neural network) we use it to represent a gradient descent step on the left value decreasing (for example) the squared difference between them. Since a maximization is required, if the network for Q were formulated as a map directly from state-action pairs to values, it would be necessary to perform one pass through the network for each action in each timestep. Because this would be terribly inefficient (particularly in the case of Hex which has up to 169 possible actions) and also because action values for a given state are highly correlated, we instead follow [10] and use a network that outputs values for all actions in one pass.

Note that since we take the maximum over the actions in each state, it is not necessary to actually follow the optimal policy to learn the optimal action values, though we do need to have some probability to take each action in the optimal policy. If the overlap with the policy followed and the optimal policy is greater we will generally learn faster. Usually the policy used is called epsilon-greedy which takes the action with the highest current $Q(s_t, a_t)$ estimate most of the time but chooses an action at random some fraction of the time. This method of exploration is far from ideal and improving on it is an interesting area of research in modern RL.

Having learned an approximation of the optimal action values, at test time we can simply pick the highest estimated action value in each state, and hopefully in doing so follow a policy that is in some sense close to optimal.

2   Method

2.1   Problem Structure

We use the convention that a win is worth a reward of $+1$ and a loss is worth -1. All moves that do not end the game are worth 0. We are in the episodic case, meaning that we wish to maximize our total reward over an episode (i.e. we want to win and not lose), hence we use no discounting ($\gamma = 1$). Note that the ground truth $q_*(S, A)$ value is either 1 or -1 for every possible state-action pair (the game is deterministic and no draws are possible, hence assuming perfect play one player is certain to lose and the other is certain to win). The network's estimated $q_*(S, A)$ value $Q(S)[A]$ then has the interpretation of subjective probability that a particular move is a win, minus the subjective probability that it is a loss (roughly speaking $Q(S)[A] = P(win|S, A) - P(loss|S, A)$). We seek to predict the true value of $q_*(S, A)$ as accurately as possible over the states encountered in ordinary play, and in doing so we hope to achieve strong play at test time by following the policy which takes action $argmax\limits_{a} Q(s)[a]$, i.e. the move with highest estimated win probability, in each state.

2.2   State Representation

The state of the Hex board is encoded as a 3 dimensional array with 2 spatial dimensions and 6 channels as follows: white stone present; black stone present; white stone group connected to left edge; white stone group connected to right edge; black stone group connected to top edge; black stone group connected to bottom edge.

In addition to the 13×13 Hex board, the input includes 2 cells of padding on each side which are connected to the corresponding edge by default and belong to the player who is trying to connect to that edge. This padding serves a dual purpose of telling the network where the edges are located and allowing both 3×3 and 5×5 filters to be placed directly on the board edge without going out of bounds. We note that AlphaGo used a far more feature rich input representation including a good deal of tactical information specific to the game of Go. Augmenting our input representation with additional information of this kind for Hex could be an interesting area of future investigation. Our input format is visualized in Fig. 3.

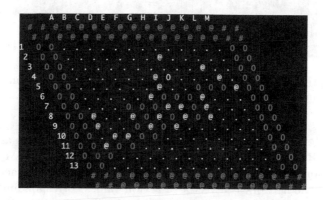

Fig. 3. A visualization of the board representation fed to NeuroHex. O is white, Ⓐ is black, red is north or east edge connected depending on the color of the associated stone, similarly green is south or west edge connected. Note that though the board is actually 13 × 13 the input size is 17 × 17 to include 2 layers of padding to represent the board edge and allow placement of 5 × 5 filters along it. Cells in the corners marked with # are uniquely colored both white and black from the perspective of the network. (Color figure online)

2.3 Model

We use the Theano library [4,6] to build and train our network. Our network architecture is inspired by that used by Google DeepMind for AlphaGo's policy network [13]. Our network consists of 10 convolutional layers followed by one fully connected output layer. The AlphaGo architecture was fully convolutional, with no fully connected layers at all in their policy network, although the final layer uses a 1 × 1 convolution. We however decided to employ one fully connected layer, as we suspect that this architecture will work better for a game in which a global property (i.e., in Hex, have you connected your two sides?) matters more than the sum of many local properties (i.e., in Go, which local battles have you won?). For future work, it would of interest to explore the effect of the final fully connected layer in our architecture (Fig. 4).

Filters used are hexagonal rather than square to better capture the different notion of locality in the game of Hex. Hexagonal filters were produced simply by zeroing out appropriate elements of standard square filters and applying Theano's standard convolution operation (see Fig. 5). Each convolutional layer has a total of 128 filters which consist of a mixture of diameter 3 and diameter 5 hexagons, all outputs use stride 1 and are 0 padded up to the size of the original input. The output of each convolutional layer is simply the concatenation of the

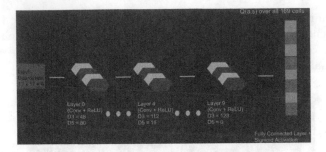

Fig. 4. Diagram showing our network layout: D3 indicates number of diameter 3 filters, D5 indicates number of diameter 5 filters in the layers shown.

padded diameter 5 and diameter 3 outputs. All activation function are Rectified Linear Units (ReLU) with the exception of the output which uses $1 - 2\sigma(x)$ (a sigmoid function) in order to obtain the correct range of possible action values. The output of the network is a vector of action values corresponding to each of the board locations. Unplayable moves (occupied cells) are still evaluated by the network but simply ignored where relevant since they are trivially pruned.

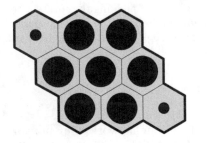
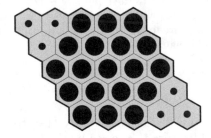

Fig. 5. Creating hexagonal filters from square filters: smaller dots are effectively ignored by fixing their weight to zero.

2.4   Position Database

While it would be possible to train the network purely by Q-learning from self-play starting from the empty board every game, we instead generated a database of starting positions from 10,000 games played by a noisy version of a strong Hex playing program based on alpha-beta search called Wolve [2]. To generate this database each game was started with a random move and each subsequent move was chosen by a softmax over Wolve's move evaluations to add additional variability. These positions were used for two separate purposes. First they were used in mentoring (supervised learning to provide a reasonable initialization of the network before Q-learning) which is described in the section below. Second to randomly draw a starting position for each episode of self-play by the network.

This second usage was meant to ensure that the network experiences a wide variety of plausible game positions during training, without going through the potentially arduous process of finding this variety by random exploration. In the future it could be interesting to see how important these two things are to the success of the algorithm.

2.5   Mentoring

Before beginning Q-learning, the network was trained to replicate (by stochastic gradient descent on the mean squared error) the action values produced by a variant of a common Hex heuristic based on electrical resistance [1], over the position database discussed in the previous section. The idea of the heuristic is to place a voltage drop across the two edges a player is trying to connect, then take the players own cells to be perfect conductors, opponent cells to be perfect insulators, and empty cells to be finite resistors. The usual version of the heuristic then computes a score (an arbitrary positive real number with no statistical interpretation) of the position as the ratio of current traveling across the board for each player. Because we wanted instead to generate heuristic action values between -1 and 1 for each move, it was necessary to modify this heuristic. We did this by computing estimates of the current across the board $C'_1(a)$ and $C'_2(a)$ for the player to move and their opponent respectively following the player to move playing into cell a (the true value could have been used by playing each move and recomputing the current, but we use a simple estimate based on the current through each cell to save time). The action value of a cell was then taken to be:

$$Q(a) = \begin{cases} 1 - C'_2(a)/C'_1(a), & \text{if } C'_1(a) > C'_2(a) \\ C'_1(a)/C'_2(a) - 1, & \text{if } C'_2(a) > C'_1(a) \end{cases}$$

In any case the details here are not terrible important and similar results could have likely been obtained with a simpler heuristic. The important thing is that the heuristic supervised pre-training gives the network some rough initial notion that the goal of the game is to connect the two sides. This serves the primary purpose of skipping the potentially very long period of training time where most updates are meaningless since the reward signal is only available at the end of an episode and every other update is mostly just backing up randomly initialized weights. It also presumably gives the network an initial set of filters which have some value in reasoning about the game. Note that the final trained network is much stronger than this initialization heuristic.

2.6   Q-learning Algorithm

We use Deep Q-learning with experience replay in a manner similar to Google DeepMind's Atari learning program [10]. Experience replay means that instead of simply performing one update at a time based on the last experience, we save a large set of the most recent experiences (in our case 100,000), and perform a random batch update (batch size 64) drawn from that set. This has a number of benefits including better data efficiency, since each experience is sampled many times and each update uses many experiences; and less correlation among sequential updates. We use RMSProp [16] as our gradient descent method.

One notable difference between our method and [10] is in the computation of the target value for each Q-update. Since in the Atari environment they have an agent interacting with an environment (an Atari game) rather than an adversary they use the update rule $Q(s_t, a_t) \xleftarrow{\alpha} r_{t+1} + \gamma \max_a Q(s_{t+1}, a)$, where again we use $\xleftarrow{\alpha}$ to indicate the network output on the left is moved toward the target on the right by a gradient descent update to reduce the squared difference. Here γ is a discount factor between 0 and 1 indicating how much we care about immediate reward vs. long-term reward.

In our case the agent interacts with an adversary who chooses the action taken in every second state, we use the following gradient descent update rule: $Q(s_t, a_t) \xleftarrow{\alpha} r_{t+1} - \max_a Q(s_{t+1}, a)$. Note that we take the value to always be given from the perspective of the player to move. Thus the given update rule corresponds to stepping the value of the chosen move toward the negation of the value of the opponents next state (plus a reward, nonzero in this case only if the action ends the game). This update rule works because with the way our reward is defined the game is zero-sum, thus the value of a state to our opponent must be precisely the negation of the value of that state to us. Also in our case we are not concerned with how many moves it takes to win and we suspect using a discount factor would only serve to muddy the reward signal so we set $\gamma = 1$.

Our network is trained only to play as white, to play as black we simply transform the state into an equivalent one with white to play by transposing the board and swapping the role of the colors. We did it this way so that the network would have less to learn and could make better use of its capacity. An alternative scheme like outputting moves for both white and black in each state seems wasteful as playing as either color is functionally the same (ignoring conventional choices of who plays first). It is however an interesting question whether training the network to pick moves for each color could provide some useful regularization.

Our Q-learning algorithm is shown in Algorithm 1. Note that we include some forms of data augmentation in the form of randomly flipping initial states to ones that are game theoretically equivalent by symmetry, as well as randomly choosing who is to move first for the initial state (irrespective of the true player to move for a given position). The latter augmentation will result in some significantly imbalanced positions since each move can be crucial in Hex and losing a move will generally be devastating. However since our player is starting off with very little knowledge of the game, having these imbalanced positions where one player has the opportunity to exploit serious weakness presumably allows the network to make early progress in identifying simple strategies. A form of curriculum learning where these easier positions are trained on first followed by more balanced positions later could be useful, but we did not investigate this here. We also flip the state resulting from each move to an equivalent state with 50% probability, a form of game specific regularization to capture a symmetry of the game and help smooth out any orientation specific noise generated in training.

initialize replay memory M, Q-network Q, and state set D
for *desired number of games* **do**
  s = position drawn from D
  randomly choose who moves first
  randomly flip s with 50% probability
  while *game is not over* **do**
    a = epsilon_greedy_policy(s, Q)
    s_{next} = s.play(a)
    if *game is over* **then**
      | $r=1$
    else
      | $r=0$
    end
    randomly flip s_{next} with 50% probability
    M.add_entry((s,a,r,s_{next}))
    $(s_t,a_t,r_{t+1},s_{t+1})$ = M.sample_batch()
    $target_t$ = $r_{t+1} - \max_a Q(s_{t+1})[a]$
    Perform gradient descent step on Q to reduce $(Q(s_t)[a_t] - target_t)^2$
    s = s.play(a)
  end
end

Algorithm 1: Our Deep Q-learning algorithm for Hex. epsilon_greedy_policy(s, Q) picks the action with the highest Q value in s 90% of the time and 10% of the time takes a random action to facilitate exploration. M.sample_batch() randomly draws a mini-batch from the replay memory. Note that in two places we flip states (rotate the board 180°) at random to capture the symmetry of the game and mitigate any orientation bias in the starting positions.

3 Results

To measure the effectiveness of our approach, we measure NeuroHex's playing strength, rate of learning, and stability of learning. Our results are summarized in Figs. 6, 9, and 10, respectively.

Figure 9 shows the average magnitude of maximal action value output by the network. Figure 10 shows the average cost for each Q-update performed during training as a function of episode number. Both of these are internal measures, indicating the reasonable convergence and stability of the training procedure; they do not however say anything about the success of the process in terms of actually producing a strong player. After an initial period of rapid improvement, visible in each plot, learning appears to become slower. Interestingly the cost plot seems to show a step-wise pattern, with repeated plateaus followed by sudden

first move	MoHex sec/move	NeuroHex black	NeuroHex white
unrestricted	1	.20	.02
all 169 openings	1	.11	.05
all 169 openings	3	.09	.02
all 169 openings	9	.07	.01
all 169 openings	30	.00	.00

Fig. 6. NeuroHex vs. MoHex win rates. Black is 1st-player. The unrestricted win-rates are over 1000 games, the others are over 169 games.

drops. This is likely indicative of the effect of random exploration, the network converges to a set of evaluations that are locally stable, until it manages to stumble upon, through random moves or simple the right sequence of random batch updates, some key feature that allows it to make further improvement. Training of the final network discussed here proceeded for around 60,000 episodes on a GTX 970 requiring a wall clock time of roughly 2 weeks. Note that there is no clear indication in the included plots that further improvement is not possible, it is simply a matter of limited time and diminishing returns.

We evaluate NeuroHex by testing against the Monte-Carlo tree search player MoHex [3,9], currently the world's strongest hexbot. See Fig. 6. Four sample games are shown in Figs. 7 and 8. On board sizes up to at least 13×13 there is a significant first-player advantage. To mitigate this, games are commonly played with a "swap rule" whereby after the first player moves the second player can elect either to swap places with the first player by taking that move and color or to continue play normally. Here, we mitigate this first-player advantage by running two kinds of tournaments: in one, in each game the first move is unrestricted; in the other, we have rounds of 169 games, where the 169 first moves cover all the moves of the board. As expected, the all-openings win-rates lie between the unrestricted 1st-player and 2nd-player win-rates. To test how win-rate varies with MoHex move search time, we ran the all-openings experiment with 1 s, 3 s, 9 s, 30 s times.

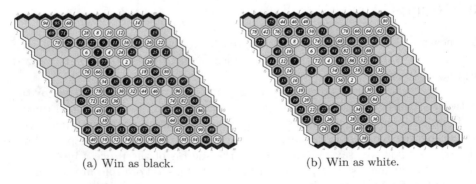

(a) Win as black.   (b) Win as white.

Fig. 7. Example wins for NeuroHex over MoHex.

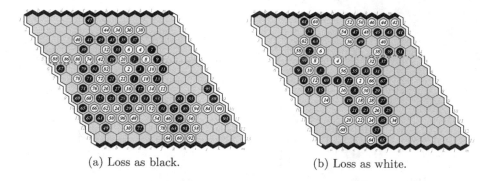

(a) Loss as black.   (b) Loss as white.

Fig. 8. Example wins for MoHex over NeuroHex.

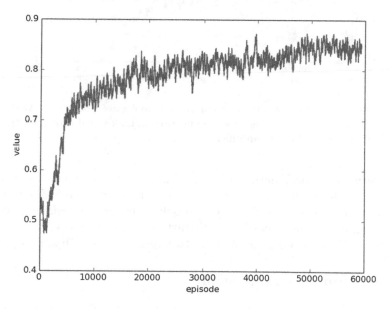

Fig. 9. Running mean (over 200 episodes) of the magnitude (absolute value) of the value (max over all action values from that position) for positions encountered in training. For each position, its ground truth value is either -1 or 1, so this graph indicates the network's confidence in its evaluation of positions that it encounters in training.

MoHex is a highly optimized C++ program. Also, in addition to Monte Carlo tree search, it uses many theorems for move pruning and early win detection. So the fact that NeuroHex, with no search, achieves a non-zero success rate against MoHex, even when MoHex plays first, is remarkable.

By comparison, AlphaGo tested their policy network against the strong Go program Pachi [5] and achieved an 85% win-rate. In this test Pachi was allowed 100,000 simulations, which is comparable to MoHex given 30 s, against which we won no games. This comparison holds limited meaning since the smaller board—

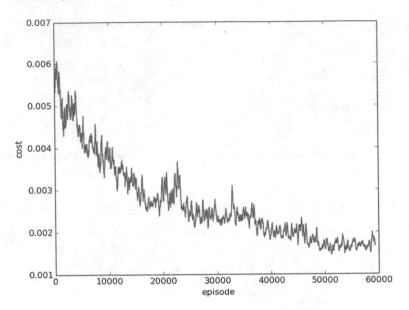

Fig. 10. Running mean (over 200 episodes) of average cost of updates in Algorithm 1: squared difference between current Q-value and target (max Q-value of next position). So this graph indicates the rate at which the network is learning.

and aggressive move pruning which is possible in Hex, unlike in Go—enhances the utility of each simulation. Nonetheless it is fair to say NeuroHex has not yet reached a comparable level to the AlphaGo policy network. It remains to be seen how much improvement is possible with further training, and to what extend our learned Q-value network can be used to improve a search based player.

4   Discussion and Conclusion

The DeepMind authors showed that Deep RL based on final outcomes can be used to build a superhuman Gobot; we have shown that Q-learning can be used to build a strong Hexbot. Go and Hex have many similarities—two-player alternate turn game on a planar board in which connectivity is a key feature—and we expect that our methods will apply to Go and many other similar games.

Before our work we wondered whether the large action spaces that arise in Hex would result in the temporal difference updates of Q-learning being over-whelmed by noise. We also wondered whether the assignment of correct credit for actions would be feasible in light of the paucity of rewards among the large number of states. But the win-rate of NeuroHex against the expert-level player MoHex after training only two weeks suggests that the generalization ability of convolutional neural networks together with the use of supervised mentor-ing is sufficient to overcome these challenges. One property of Hex that might have contributed to the success of our methods is that—unlike Go—it is an all-or-nothing game. In particular, Hex has a "sudden death" property: in many positions, for most moves, it is easy to learn that those moves lose. In such positions it is comparatively easy task for the network to learn to distinguish the few good moves from the many bad ones.

In light of supervised mentoring one could ask to what extent our training is using the reward signal at all, versus simply back-propagating the heuristic initialization values. We would like to address this question in the future, for example by testing the procedure without supervised mentoring, although this might not be important from the perspective of building a working system. If the heuristic is good then many of the values should already be close to the true value they would eventually converge to in Q-learning. Assuming, as is often the case, that heuristic values near the endgame are better than those near the start, we will be able to perform meaningful backups without directly using the reward signal. To the extent the heuristic is incorrect it will eventually be washed out by the reward signal—the true outcome of the game—although this may take a long time.

We suspect that our network would show further improvement with further training, although we have no idea to what extent. We also suspect that incorporating our network into a player such as MoHex, for example to bias the initial tree search, would strengthen the player.

5   Future Work

Throughout the paper we have touched on possible directions for further research. Here are some possibilities: augment the input space with tactical or strategic features (e.g. in Hex, virtual connections, dead cells and capture patterns); build a search based player using the trained network, or incorporate it into an existing search player (e.g. MoHex); determine the effectiveness of using an actor-critic method to train a policy network alongside the Q network to test the limits of the learning process; find the limits of our approach by training for a longer period of time; determine whether it is necessary to draw initial positions from a state database, or to initialize with supervised learning; investigate how using a fully convolutional neural network compares to the network with one fully connected layer we used.

References

1. Anshelevich, V.V.: The game of Hex: an automatic theorem proving approach to game programming. In: AAAI/IAAI, pp. 189–194 (2000)
2. Arneson, B., Hayward, R., Henderson, P.: Wolve wins Hex tournament. ICGA J. **32**, 49–53 (2008)
3. Arneson, B., Hayward, R.B., Henderson, P.: Monte Carlo tree search in Hex. IEEE Trans. Comput. Intell. AI Games **2**(4), 251–258 (2010)
4. Bastien, F., Lamblin, P., Pascanu, R., Bergstra, J., Goodfellow, I.J., Bergeron, A., Bouchard, N., Bengio, Y.: Theano: new features and speed improvements. In: NIPS 2012 Deep Learning and Unsupervised Feature Learning Workshop (2012)

5. Baudiš, P., Gailly, J.: PACHI: state of the art open source go program. In: Herik, H.J., Plaat, A. (eds.) ACG 2011. LNCS, vol. 7168, pp. 24–38. Springer, Heidelberg (2012). doi:10.1007/978-3-642-31866-5_3
6. Bergstra, J., Breuleux, O., Bastien, F., Lamblin, P., Pascanu, R., Desjardins, G., Turian, J., Warde-Farley, D., Bengio, Y.: Theano: a CPU and GPU math expression compiler. In: Proceedings of the Python for Scientific Computing Conference (SciPy) (2010). Oral Presentation
7. Gardner, M.: Mathematical games. Sci. Am. **197**(1), 145–150 (1957)
8. Hayward, R.B.: MoHex wins Hex tournament. ICGA J. **36**(3), 180–183 (2013)
9. Huang, S.-C., Arneson, B., Hayward, R.B., Müller, M., Pawlewicz, J.: MoHex 2.0: a pattern-based MCTS Hex player. In: Herik, H.J., Iida, H., Plaat, A. (eds.) CG 2013. LNCS, vol. 8427, pp. 60–71. Springer, Cham (2014). doi:10.1007/978-3-319-09165-5_6
10. Mnih, V., Kavukcuoglu, K., Silver, D., Rusu, A.A., Veness, J., Bellemare, M.G., Graves, A., Riedmiller, M., Fidjeland, A.K., Ostrovski, G., Petersen, S., Beattie, C., Sadik, A., Antonoglou, I., King, H., Kumaran, D., Wierstra, D., Legg, S., Hassabis, D.: Human-level control through deep reinforcement learning. Nature **518**(7540), 529–533 (2015)
11. Reisch, S.: Hex ist PSPACE-vollständig. Acta Informatica **15**, 167–191 (1981)
12. Shannon, C.E.: Computers and automata. Proc. Inst. Radio Eng. **41**, 1234–1241 (1953)
13. Silver, D., Huang, A., Maddison, C.J., Guez, A., Sifre, L., van den Driessche, G., Schrittwieser, J., Antonoglou, I., Panneershelvam, V., Lanctot, M., Dieleman, S., Grewe, D., Nham, J., Kalchbrenner, N., Sutskever, I., Lillicrap, T., Leach, M., Kavukcuoglu, K., Graepel, T., Hassabis, D.: Mastering the game of Go with deep neural networks and tree search. Nature **529**(7587), 484–489 (2016)
14. Sutton, R.S., Barto, A.G.: Reinforcement Learning: An Introduction. MIT Press, Cambridge (1998)
15. Tesauro, G.: Temporal difference learning and TD-gammon. Commun. ACM **38**(3), 58–68 (1995)
16. Tieleman, T., Hinton, G.: Lecture 6.5–RmsProp: divide the gradient by a running average of its recent magnitude. COURSERA Neural Netw. Mach. Learn. (2012)

Deep or Wide? Learning Policy and Value Neural Networks for Combinatorial Games

Stefan Edelkamp[✉]

Faculty of Mathematics and Computer Science,
University of Bremen, Bremen, Germany
edelkamp@tzi.de

Abstract. The success in learning how to play *Go* at a professional level is based on training a deep neural network on a wider selection of human expert games and raises the question on the availability, the limits, and the possibilities of this technique for other combinatorial games, especially when there is a lack of access to a larger body of additional expert knowledge.

As a step towards this direction, we trained a *value network* for Tic-TacToe, providing perfect winning information obtained by *retrograde analysis*. Next, we trained a *policy network* for the SameGame, a challenging combinatorial puzzle. Here, we discuss the interplay of deep learning with *nested rollout policy adaptation* (NRPA), a randomized algorithm for optimizing the outcome of single-player games.

In both cases we observed that ordinary feed-forward neural networks can perform better than convolutional ones both in accuracy and efficiency.

1 Introduction

Deep Learning[1] is an area of AI research, which has been introduced with the objective of moving the field closer to its roots: the creation of human-like intelligence. One of its core data structures is a convolutional neural network (CNN). As in conventional NNs, CNNs are trained with reinforcement learning (back-propagation and stochastic gradient decent). The major advance in learning larger NNs are growing resources in computational power, especially in graphics processing units (GPUs) found on the computer's graphics card.

Prototypical applications for deep learning are Computer Vision, but also Language Understanding. In Game Playing, deep learning has made its way to play real-time strategy games by just looking at the screen and score data [6]. More formally the aim of this reinforcement technique is to learn a *policy network*, which outputs a probability distribution of the next move to play. Alternatively, in a *value network*, learning is used to predict the *game-theoretical value* in a single network output node, i.e., its expected outcome assuming perfect play.

---

[1] http://deeplearning.net.

© Springer International Publishing AG 2017
T. Cazenave et al. (Eds.): CGW 2016/GIGA 2016, CCIS 705, pp. 19–33, 2017.
DOI: 10.1007/978-3-319-57969-6_2

As a sign of industrial relevance, Google bought the deep learning specialist DeepMind. DeepMind's *AlphaGo*[2] is a Go game playing program that applies a combination of neural network learning and Monte Carlo tree search. In March 2016, it won 4:1 against Lee Sedol in a match; proving itself to be the first computer program to ever beat a professional human player in Go [9]. This achievement is widely considered a landmark result in AI, previously estimated to become true only in the far future. The program *AlphaGo* was trained both with games played by humans and with ones played by earlier versions of *AlphaGo*. It is amazing that deep learning of thousands of expert games (in matters of days of GPU computation time) made the program *understand* the strategic concepts and tactics of Go.

AlphaGo learned to match the moves of expert players from recorded historical games. Once it had reached a certain degree of proficiency, it was trained further by playing games against other instances of itself. The input was a random permutation of expert game positions, made available in a number of Boolean input matrices of size 19×19 (some for the occupation and the colors that play, some for further features like liberty). The output was 19×19 matrix as a predictor for the next move. The convolutional neural network (CNN) that was trained was a *policy network*.

Cazenave [3] could reproduce *AlphaGo*'s results by training a deep neural network for the same set of expert games on a GPU-lifted PC infrastructure. The minimized error evaluation was comparable to the one obtained and reported by DeepMind [9]. The accuracy of finding the correct expert move was 55.56%, while *AlphaGo*had success rate of 57.0%.

Collocated to a human top tournament (79th Tata Steels) for the Google Alpha Chess Classics in Wijk an Zee 2017[3] an entire tournament was set, where Alpha-Go-inspired chess engines will try playing close to the humans from whose games they has been trained, e.g., Anderssen, Pillsbury, Tarrasch, Capablanca, Tal, Smyslov, Fischer, some of which have not played against each other in real live.

Following Rojas [7], a neural network is graph representation of a function with input variables in $\mathbb{R}^l$ and output variables in $\mathbb{R}^k$. The internal working is described through an activation function and a threshold applied at each network node. The input vector is a number of l features (e.g., in board game the board itself is often included in the feature vector). In a value network we have $k = 1$, while in policy networks we get a probability distribution for the successor set. Learning is the process of computing weights to the network edges to find a close approximation of the true network function via the (incremental) change of the weights through labeled input/output sample pairs. In *multi-layer feed-forward neural networks* (MLNN) there are input and output, as well as fully connected intermediate hidden layers, while for CNNs the input layers are more loosely connected through a number of *planes*.

---

[2] https://deepmind.com/alpha-go.html.
[3] See de.chessbase.com/post/google-alpha-chess-classic.

The rest of the paper is structured as follows. First we introduce the concept learning of games and take TicTacToe as a first case study. We explain the steps taken to a trained neural network that can be used as an evaluation function to play the game. By the small problem size, this part of the paper has a tutorial character to guide the reader through the practical steps needed to construct a neural network game player. Nonetheless, the learning results are interesting, in the sense that the CNNs had problems to match the efficiency of MLNNs. Next, we turn to the SameGame, a single-agent maximization problem. We provide a state-of-the art NRPA player implementation of the game to compute the training sets for the neural network. Results of training CNNs and MLNNs are shown and discussed.

2   Case Study: TicTacToe

We exemplify the learning setup in TicTacToe (Fig. 1), where we construct and train a *value network*. The game is a classic puzzle that results in a draw in optimal play[4].

```
X _ X        1 0 1   1 0 1   0 0 0
_ O _   ->   0 1 0   0 0 0   0 1 0
O _ X        1 0 1   0 0 1   1 0 0
```

Fig. 1. A TicTacToe position won for the X player, and its representation in form of input planes.

We used the prominent tensor and optimization framework *torch7*, which provides an interactive interface for the programming language LUA[5]. *Tensors* featured by the programming framework are numerical matrices of (potentially) high dimension. It already offers the support for optimizers like stochastic gradient decent, as well as neural network designs and training. For fast execution of tensor operations, *torch7* supports the export of computation to the graphic card (GPU) in CUDA[6], a GPU programming framework that is semantically close to and finally links to C. The changes to the LUA code are minimal.

2.1   Automated Generation of Games

We kicked off with generating all 5478 valid TicTacToe positions, and determined their true game value by applying *retrograde analysis*, a known technique for constructing strong solutions to games. The according code is shown in Fig. 2. All classified TicTacToe positions are stored in *comma separated value* (CSV) files.

---

[4] This has lead movies like *war games* to use it as an example of a game that consumes unlimited compute power to solve.

[5] http://torch.ch/, the Linux installation is simple if the firewall does not block Githup access.

[6] https://developer.nvidia.com/cuda-zone.

```
retrograde()
  change = 1
  while (change)
    change = 0
    for each c = 1 .. 5478
      if (solved[c] == UNSOLVED)
        unpack(c)
        succs = successors(moves)
        if (moveX())
          onesuccwon = 0;
          allsuccslost = 1
          for each i = 1 .. succs
            apply(moves[i],'X')
            onesuccwon |=
              (solved[pack()] == WON)
            allsuccslost &= (
              solved[pack()] == LOST)
            apply(moves[i],'_')
          if (succs & onesuccwon)
            solved[c] = WON; change = 1
          if (succs && allsuccslost)
            solved[c] = LOST; change = 1
        else
          onesucclost = 0;
          allsuccswon = 1
          for each i = 1 .. succs
            apply(moves[i],'O')
            onesucclost |=
              (solved[pack()] == LOST)
            allsuccswon &=
              (solved[pack()] == WON)
            apply(moves[i],'_')
          if (succs && onesucclost)
            solved[c] = LOST;
            change = 1
          if (succs && allsuccswon)
            solved[c] = WON;
            change = 1
  for each c = 1 .. 5478
    if (solved[c] == UNSOLVED)
      solved[c] = DRAW
```

Fig. 2. Finding the winning sets in TicTacToe.

In one *network output* file the *values* for the value network are kept (for a policy network a suitable policy has to be used). In the other *network input* file, we recorded the according 5478 intermediate game positions. For each position, we took three 3×3 Boolean *planes* to represent the different, boards, one for the free cells, one for the X player and one for the O player.

2.2 Defining the Network

Next, we produced the input and output files for the neural network to be trained and tested. As shown in Fig. 3 we used *torch7* for the compilation of entries from the CSV input to the required binary format.

The NN consists of layers that are either fully connected (MLNN) or convoluted (CNN). The according LUA code is shown in Figs. 4 and 5. For CNNs it consists of a particular layered structure, which is interconnected through the

```
local Planes = 3
local csvFile = io.open('ttt-input.csv','r')
local input = torch.Tensor(5478,nPlanes,3,3)
local nb = 0
local currentnb = 0
for line in csvFile:lines('*l') do
  nb = nb + 1
  currentnb = currentnb + 1
  local l = line:split(',')
  local plane = 1
  local x = 1
  local y = 1
  for key, val in ipairs(l) do
    input[currentnb][plane][x][y] = val
    y = y + 1
    if y == 4 then
      y = 1
      x = x + 1
    end
    if x == 4 then
      x = 1
      plane = plane + 1
    end
  end
  if currentnb == 5478 then
    currentnb = 0
    nameInputFile = 'ttt-input.dat'
    torch.save (nameInputFile, input)
  end
  if nb == 5478 then
    break
  end
end
csvFile:close()
```

Fig. 3. Converting TicTacToe CSV to a tensor file.

```
require 'nn'
local net = nn.Sequential ()
net:add (nn.Reshape(27))
net:add (nn.Linear(27,512))
net:add (nn.Tanh())
net:add (nn.Linear(512,1))
local nbExamples = 5478
local input = torch.load ('ttt-input.dat')
local output = torch.load ('ttt-output.dat')
dataset = {};
function dataset:size() return nbExamples end
for j = 1, dataset:size() do
  dataset[j] = {input[j], output[j]};
end
criterion = nn.MSECriterion()
trainer = nn.StochasticGradient(net,criterion)
trainer.maxIteration = 1500
trainer.learningRate = 0.00005
trainer:train(dataset)
```

Fig. 4. Learning TicTacToe with an MLNN.

```
require 'nn'
local nPlanesInput = 3
local net = nn.Sequential ()
local nplanes = 25
net:add (nn.SpatialConvolution
  (nPlanesInput, nplanes, 3, 3, 1, 1, 0, 0))
net:add (nn.ReLU ())
net:add (nn.SpatialConvolution
  (nplanes, nplanes, 2, 2, 1, 1, 1, 1))
net:add (nn.ReLU ())
net:add (nn.SpatialConvolution
  (nplanes, nplanes, 2, 2, 1, 1, 1, 1))
net:add (nn.ReLU ())
net:add (nn.SpatialConvolution
  (nplanes, 1, 3, 3, 1, 1, 1, 1))
net:add (nn.ReLU ())
print(net)
net:add (nn.Reshape(1*3*3))
net:add (nn.Linear(9,1))
local nbExamples = 5478
local input = torch.load ('ttt-input.dat')
local output = torch.load ('ttt-output.dat')
dataset = {};
function dataset:size() return nbExamples end
for j = 1, dataset:size() do
  dataset[j] = {input[j], output[j]};
end
criterion = nn.MSECriterion()
trainer = nn.StochasticGradient(net,criterion)
trainer.maxIteration = 1500
trainer.learningRate = 0.00005
trainer:train(dataset)
```

Fig. 5. Learning to play TicTacToe with a CNN.

definition of planes in form of tensors. The hidden units were automatically generated by the tensor dimensions. This was done though defining sub-matrices of certain sizes and some padding added to the border of the planes. After having the input planes T_I represented as tensors and the output planes represented as tensors T_O (in our case a singular value) there are $k - 2$ spatial convolutions connected by the tensors $T_I = T_1, \ldots, T_k = T_O$. The information on the size of sub-matrices used and on the padding to the matrix was used as follows. All possible sub-matrices of a matrix for a plane (possibly extended with the padding) on both sides of the input are generated. The sub-matrices are fully connected and the matrices themselves.

2.3   Training the Network

Deep learning in CNNs is very similar to learning in classical NNs. The main exception is an imposed particular network structure and the computational power to train even larger networks to a small error. For global optimization, usually stochastic gradient decent is used [1].

Figures 4 and 5 also show the LUA code for training the network. We experimented with an alternative formulation for the optimization process, but while other NN experts insist on batched learning to be more effective, for us it did not made much of a difference.

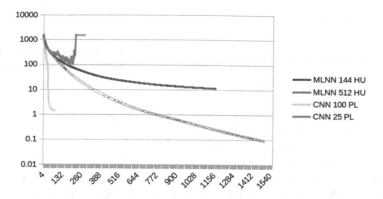

Fig. 6. Learning results in TicTacToe displaying the training error for the full set with multi-layer neural nets (MLNN) and convolutional neural nets (CNN).

Figure 6 shows the effect of learning different NNs design given the precomputed classification of all valid TicTacToe positions. We see that larger networks (number of hidden units - HU, number of intermediate planes - PL) yield better learning. Moreover, CNNs tend to have the smaller number of learning epochs CNNs compared to MLNNs. However, each optimization step in a CNN is considerably slower than in a MLNN[7]. While learning MLNNs yields a rather smooth monotone decreasing curve the learning in a CNN has more irregularities. Moreover, CNNs tend to saturate. Worse, we observe that after reaching some sweet spot CNNs can even deviate back to very bad solutions.

2.4   Using the Network

A trained value network can be used as an estimator of game positions (usually called evaluation function) and integrated in other game playing programs. As TicTacToe is a known draw we are more interested in the average value accuracy in training vs. test data.

The following small test shows that the network has learned the value of the game for which we chose 0 for a won game, 50 for draw, and 100 for a lost game. The value network it can be used as an evaluation function and, thus, immediately leads to playing engines. In fact, by fixing the class ranges to 25 and 75, we could use the trained net as an optimal predictor of the true winning classes.

---

[7] All experiments are executed one core of an Intel® Core™ i5-2520M CPU @ 2.50 GHz × 4. The computer has 8 GB of RAM but all invocations of the algorithm to any problem instance used less than 1 GB of main memory. Moreover, we had the following software infrastructure. Torch7 with LUA, Operating system: Ubuntu 14.04 LTS, Linux kernel: 3.13.0-74-generic.

3   Case Study: SameGame

The SameGame is an exciting interactive game for humans and for computers played on an $n \times n$ (usually, $n = 15$ and $k = 5$) board with k colored tiles. Tiles can be removed, if they form a connected group of $l > 1$ elements. The score of a move is $(l - 2)^2$ points. If a group of tiles is removed, others fall down. If a column becomes empty, others move to the left, so that all non-empty columns are aligned. Total clearance yields an additional bonus of 1,000 points. The objective is to maximize the score.

Successor generation and evaluation has to find the tiles that have the same color. Figure 9 shows a simplified implementation for generating the successors. We used an explicit stack for building moves. Termination is checked by looking into each of the 4 directions for a tile of the same color.

3.1   Randomized State-Space Search

As the access to high-quality expert games in many domains is limited, the research question addressed in this paper is how to apply (deep) learning in state-space search without the input of human knowledge, where *state-space search* is a general term for the exploration of problem domains to search a solution that optimizes a given cost function [4]. State spaces are generated if the problem are intrinsically hard, which is often the case in games, due to the inherent combinatorial structure. The enumeration of state spaces, however, often suffers from the *state-explosion problem*, which states that the sizes of the spaces are exponential in the number of state variables (Fig. 7).

Randomized algorithms often show performance advantages to deterministic algorithms, as in the randomized test for primality [10,11]. In *Roshambo*, random strategies are superior to deterministic ones. Randomization often turns out to be conceptually simple and, frequently, successful in large state-spaces to find the *needle in-the-haystack*. For example, most successful SAT solvers like *Lingeling*[8] rely on randomized search (Fig. 8).

In the domain of single-agent combinatorial games, nested rollout policy adaptation (NRPA) is a randomized optimization algorithm that has found new records for several problems [8]. Besides its excellent results in games, it

Predicted Value	True Value
95.6280	100
-3.1490	0
50.8897	50
0.6506	0
⋮	⋮

Fig. 7. Accuracy of value neural network for TicTacToe.

---

[8] http://fmv.jku.at/lingeling, with parallel implementations Plingeling & Treengeling.

```
4 2 2 5 2 1 5 1 5 5 2 2 1 3 4
4 4 3 1 5 5 2 4 2 3 1 1 5 1 5
1 3 4 5 4 1 4 1 1 4 5 5 2 2 2
3 4 5 1 3 4 1 3 5 5 5 4 1 3 4                    4
2 3 2 4 2 3 1 2 3 2 1 4 5 1 2                    5
1 5 5 4 1 4 5 3 3 3 1 3 4 5 1                    2
3 5 4 5 3 4 2 2 2 4 5 2 1 4 2                    4
2 1 1 5 1 4 2 3 2 1 5 2 4 4 2                    2
2 4 4 3 1 5 4 2 4 1 5 2 1 1 4                    4
1 4 4 5 3 4 1 1 3 2 3 4 5 1 2                    2
1 5 2 3 1 2 4 5 4 4 5 2 5 1 5                  3 5
3 3 4 2 1 5 1 2 3 5 2 4 4 1 2                2 1 2
4 4 1 3 4 3 2 5 4 2 4 1 3 2 4                3 2 4
2 1 4 3 2 5 5 5 5 1 5 3 2 4 5    1 2     5   3 2 3 5
2 1 2 1 2 2 3 3 2 1 1 2 5 4 3    2 5 1 5 3 5 2 5 2 3
```

Fig. 8. Initial and final position in the SameGame.

```
legalMoves(m[])
  succs = 0;
  visited.clear()
  if (!moreThanOneMove (tabu))
    tabu = blank;
  for each i = 1 .. n^2-1
    if (color[i] != blank)
      if (visited[i] == 0)
        buildMove (i, m[succs]);
        succs += |m[succs].tiles| > 1
  return succs

play (move)
  move.sort()
  for each i = 0 .. |move.tiles|-1
    remove(move.locations[i])
  column = 0
  for each i = 0 .. n-1
    if (color[n*(n-1) + column] == blank)
      removeColumn (column);
    else
      column++;
  currentScore += (|move.tiles|-2)^2
    + 1000 * color[n*(n-1)] == blank)
  rollout[length++] = move
```

Fig. 9. Generating the successors and executing a move in the SameGame.

has been effective in the applications, for example in IT logistics for solving
constrained routing problems, or in computational biology for computing and
optimizing multiple sequence alignments. NRPA belongs to the wider class of so-
called Monte Carlo tree search (MCTS) algorithms, where Monte Carlo stands
as an alias for random program execution. The main concept of MCTS is the
playout (or rollout) of positions, with results in turn change the likelihood of the
generation of successors in subsequent trials. Other prominent members in this
class are *upper confidence bounds applied to trees* (UCT) [5] (that was applied to
in *AlphaGo*), and *nested monte-carlo tree search* (NMCTS) that has been used
successfully for combinatorial single-agent games with large branching factor [2].
What makes NRPA different to UCT and NMCTS is the concept of training a
policy through a mapping of encoded moves to probabilities (in NMCTS the

```
Nrpa(level)
  Solution best
  if (level == 0)
    best.score = playout(global)
    best.rollout = rollout
  else
    best.score = Init
    backup[level] = global
    for(i=0..iteration)
      Solution r = Nrpa(level - 1)
      if (better(r.score,best.score))
        best = r
      adapt(r.score,r.rollout,backup[level])
    global = backup[level]
  return best
```

Fig. 10. Nested rollout policy adaptation.

policy is hidden in the recursive structure of the program's decision-making, while in UCT the policy is represented partitioning in the nodes of the top tree that is stored).

NRPA was introduced in the context of generating a new world record in *Morpion Solitaire*. While the algorithm is general and applies to many other domains, we keep the notation close to games and will talk about boards, rollouts, moves. The recursive search procedure is shown in Fig. 10. It requires a proper initialization value *Init* and comparison function *Better*, depending on whether a maximization or a minimization problem being solved. Different to UCT and NMCS, in NRPA every playout starts from an empty board. Two main parameters trade exploitation vs. exploration: the number of *levels* and the branching factor *iteration* of successors in the recursion tree. There is a training parameter α, usually kept at $\alpha = 1$. Successor selection refers to probabilities for each move are computed and recorded in a distribution vector. We implement *playout* and *adapt* based on domain-depending successor generation and move encoding rules, functions `terminal`, `legalMoves`, `code`. The function `code` maps the move to an integer that addresses the value in the policy table. As it is called only during a playout it has access to all other information of the state that is produced. This way *code* realizes a mapping of state and move to a floating-point value.

The implementation for policy adaptation in Fig. 12 records the codes and the length of the playout in the successor selection *Select* (Fig. 11). This leads to the implementation of the generic *Playout* function (Fig. 13): each time a

```
Select(board, moves, pol)
  for each i .. moves.size
    c = board.code(moves[i]);
    probaMove[i] = exp(pol[c]);
    bestCode[0][board.length][i] = c;
  return probaMove
```

Fig. 11. A fitness selection module.

```
Adapt(length, level, p)
  for each i = 0 .. length
    backup[level][bestCode[level][i]] =
      bestCode[level][i] + ALPHA
    z = 0
    for each j = 0 .. |bestCode[level][i]|
      z += exp(pol[bestCode[level][i][j]])
    for each j = 0 .. |bestCode[level][i]|
      p'[level][bestCode[level][i][j]] -=
        ALPHA * exp(p[bestCode[level][i][j])/z
```

Fig. 12. An implementation of policy adaptation.

```
Playout(pol)
  Board b
  while(1)
    if (board.terminal())
      score = board.score ()
      scoreBestRollout[0] = score
      lengthBestRollout[0] = board.length
      for each k = 0 .. board.length
        bestRollout[0][k] = board.rollout[k]
      if (Better(score,bestScore))
        bestScore = score
        bestBoard = board
      return score
    moves = board.legalMoves(moves)
    nbMovesBestRollout[0][board.length] =
      moves.size
    probaMove = Select(board,moves,pol)
    sum = probaMove[0]
    for each i = 1 .. |provaMove|
      sum += probaMove[i]
    r = rand(0,sum)
    j = 0
    s = probaMove[0]
    while (s < r)
      s += probaMove[++j]
    bestCode[0][board.length] =
      bestCode[0][board.length][j]
    board.play(moves[j])
```

Fig. 13. The generic playout function.

new problem instance in form of an initial board is created. With *Select* the procedure calls the fitness evaluation.

3.2 Generating the Training Data

As a first step we generate input data for training the CNN using our Monte Carlo tree search solver. We used the benchmark set of 20 problem instances[9] with board sizes $n = 15$. Each tile has one of 5 colors.

We ran a level-3 iteration-100 NRPA search 30 times. To compare with we also ran one NRPA(4,100) for each problem. All individual games were recorded, merged and subsequently split into 33972 partial states, one after each move, The

---

[9] http://www.js-games.de/eng/games/samegame.

partial state were stored into an input file. For each partial state the move executed was stored into another file. The 33972 partial states were chosen randomly to avoid a bias in training the network.

3.3 Defining the Network

To specify a policy network for the SameGame the set of input planes were defined as follows. For each of the 5 colors plus 1 for the blank, we defined an

Table 1. Parameter finding for deep learning in the SameGame using 1000 of 33972 randomly chosen training examples, minimizing the MSE in stochastic gradient decent according to different learning rates λ.

$\lambda = 0.0005$	$\lambda = 0.005$	$\lambda = 0.05$	$\lambda = 0.5$	$\lambda = 0.2$
0.1429	0.1548	0.1437	0.2084	0.1567
0.1409	0.1418	0.1414	0.2088	0.1537
0.1404	0.1409	0.1398	0.2088	0.1533
0.1400	0.1408	0.1392	0.2088	0.1531
0.1395	0.1408	0.1388	0.2088	0.1527
0.1391	0.1407	0.1384	0.2088	0.1523
0.1387	0.1407	0.1382	0.2088	0.1521
0.1384	0.1406	0.1380	0.2088	0.1519
0.1382	0.1406	0.1378	0.2088	0.1517
0.1380	0.1406	0.1376	0.2088	0.1515
0.1378	0.1405	0.1375	0.2088	0.1515
0.1376	0.1405	0.1374	0.2088	0.1501
0.1374	0.1405	0.1373	0.2088	0.1457
0.1372	0.1404	0.1371	0.2088	0.1423
0.1370	0.1404	0.1349	0.2088	0.1434
0.1368	0.1404	0.1320	0.2088	0.1416
0.1366	0.1403	0.1291	0.2088	0.1373
0.1365	0.1403	0.1327	0.2088	0.1359
0.1363	0.1403	0.1324	0.2088	0.1353
0.1362	0.1402	0.1325	0.2088	0.1348
0.1361	0.1402	0.1323	0.2088	0.1338
0.1360	0.1402	0.1323	0.2088	0.1333
0.1359	0.1401	0.1322	0.2088	0.1344
0.1358	0.1401	0.1322	0.2088	0.1346
0.1357	0.1401	0.1321	0.2088	0.1352
$\vdots$	$\vdots$	$\vdots$	$\vdots$	$\vdots$

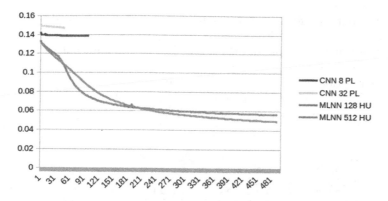

Fig. 14. Learning results in the SameGame displaying the change of the network error on the full training set of 33972 game positions for multi-layer neural nets (MLNN) and convolutional neural nets (CNN).

indicator matrix of size 15×15 for the board, denoting if a tile is present in a cell or not. This amounts to 6 planes of size 225, so that we had 1350 binary input features to the neural network. The output tensor file refers to one binary plane of size 15×15 representing a board, with the matrix entries denote the cells affected by the move. What was learned in this *policy network* is the distribution values on where to click.

3.4 Training the Network

Table 1 shows the effect of varying the learning parameter for the learning process on a fraction of all training examples. In Fig. 14 we see the effect of learning different neural networks given the precomputed randomly perturbed set of all SameGame training positions. The learning rate was 0.1–0.2 and the first 50–500 iterations of the optimization process are plotted. Again, it seemingly looks like that MLNNs can perform better in comparison with convolutional structures. Moreover the convergence was much faster, the largest MLNN took about 5 h and the smallest about 1.5 h, while the CNN took about 2 days of computational time on our CPU.

3.5 Using the Network

To validate our solution, we compared the MLNN network output after 1000 learning epochs (having an error of 0.0430463278) with the real output. In the visualization of Fig. 15 we used the threshold of 0.2 for a bit being set.

We see that more time is needed to reduce the error to a value in which can be used for playing well.

The subsequent integration of the neural network into the randomized NRPA engine, however, is simple, as we only need to change the initialization or rollout functions. There are three main implementation options.

- The distribution information is used as an initial policy matrix prior to the search.
- The NN recommendation and the learned policy are alternated by flipping a coin with probability p.
- The distribution of successors computed by the policy are merged with the NN recommended ones. If p_i and p'_i are the two probabilities for choosing the i-th of r successors, then the new probability is $p_i \cdot p'_i / \sum_{k=1}^{r} p_k \cdot p'_k$.

```
real output                    predicted output
0 0 0 0 0 0 0 0 0 0 0 0 0 0 0  0 0 0 0 0 0 0 0 0 0 0 0 0 0 0
0 0 0 0 0 0 0 0 0 0 0 0 0 0 0  0 0 0 0 0 0 0 0 0 0 0 0 0 0 0
0 0 0 0 0 0 0 0 0 0 0 0 0 0 0  0 0 0 0 0 0 0 0 0 0 0 0 0 0 0
0 0 0 0 0 0 0 0 0 0 0 0 0 0 0  0 0 0 0 0 0 0 0 0 0 0 0 0 0 0
0 0 0 0 0 0 0 0 0 0 0 0 0 0 0  0 0 0 0 0 0 0 0 0 0 0 0 0 0 0
0 0 0 0 0 0 0 0 0 0 0 0 0 0 0  0 0 0 0 0 0 0 0 0 0 0 0 0 0 0
0 0 0 0 0 0 0 0 0 0 0 0 0 0 0  0 0 0 0 0 0 0 0 0 0 0 0 0 0 0
0 0 0 0 0 0 0 0 0 0 0 0 0 0 0  0 0 0 0 0 0 0 0 0 0 0 0 0 0 0
0 0 0 0 0 0 1 0 0 0 0 0 0 0 0  0 0 0 0 0 0 0 0 0 0 0 0 0 0 0
0 0 0 0 0 1 1 1 1 1 0 0 0 0 0  0 0 0 0 0 1 1 1 1 1 0 0 0 0 0
0 0 0 0 0 0 0 0 1 0 0 0 0 0 0  0 0 0 0 0 0 0 0 1 0 0 0 0 0 0
0 0 0 0 0 0 0 1 0 0 0 0 0 0 0  0 0 0 0 0 0 0 0 0 0 0 0 0 0 0
0 0 0 0 0 0 0 0 0 0 0 0 0 0 0  0 0 0 0 0 0 0 0 0 0 0 0 0 0 0
0 0 0 0 0 0 0 0 0 0 0 0 0 0 0  0 0 0 0 0 0 0 0 0 0 0 0 0 0 0
0 0 0 0 0 0 0 0 0 0 0 0 0 0 0  0 0 0 0 0 0 0 0 0 0 0 0 0 0 0
```

Fig. 15. Validation of learning result.

4   Conclusion

This paper explains the working of (deep) learning for training value and policy (neural) networks, to reflect its usage in game playing programs. In both of our case studies, we excluded human expert knowledge and used accurate and approximate computer exploration results instead.

Deep learning for TicTacToe is like shooting large bullets on too small animals, especially given that we have computed exact information on the game theoretical value for all reachable states beforehand. Nonetheless, we see the results of the learning process as being insightful. We were able to train the network to eventually learn the exact winning information in TicTacToe, and likely due to better separation, the wider the hidden layer(s), the better the learning.

Next, we turned to the SameGame, for which we applied a fast randomized solver. We used it to generate a series of good games (30 for each of the considered 20 instances). We obtained better learning curve with shallow MLNNs, which also lead to a drastic performance gain (about 20–30 fold speedup) compared to our CNN designs.

The sparser form of convolutions are often reported to perform better than ordinary multi-layerd neural networks that have fully connected hidden layers. The sparser interconnection between the network levels is balanced by a deeper network. To some extend our results can be interpreted in the sense that good neural learning does not always has to be *deep*, but sometimes *wide* and *shallow*.

References

1. Bottou, L.: Stochastic learning. In: Bousquet, O., Luxburg, U., Rätsch, G. (eds.) ML -2003. LNCS (LNAI), vol. 3176, pp. 146–168. Springer, Heidelberg (2004). doi:10.1007/978-3-540-28650-9_7
2. Cazenave, T.: Nested Monte-Carlo Search. In: IJCAI, pp. 456–461 (2009)
3. Cazenave, T.: Combining tactical search and deep learning in the game of go. In: IJCAI-Workshop on Deep Learning for Artificial Intelligence (DLAI) (2016)
4. Edelkamp, S., Schrödl, S.: Heuristic Search - Theory and Applications. Academic Press, London (2012)
5. Kocsis, L., Szepesvári, C.: Bandit based Monte-Carlo planning. In: ECML, pp. 282–293 (2006)
6. Mnih, V., Kavukcuoglu, K., Silver, D., Rusu, A.A., Veness, J., Bellemare, M.G., Graves, A., Riedmiller, M., Fidjeland, A.K., Ostrovski, G., Petersen, S., Beattie, C., Sadik, A., Antonoglou, I., King, H., Kumaran, D., Wierstra, D., Legg, S., Hassabis, D.: Human-level control through deep reinforcement learning. Nature 518(7540), 529–533 (2015)
7. Rojas, R.: Neural Networks: A Systematic Introduction. Springer, New York (1996)
8. Rosin, C.D.: Nested rollout policy adaptation for Monte-Carlo tree search. In: IJCAI, pp. 649–654 (2011)
9. Silver, D., Huang, A., Maddison, C.J., Guez, A., Sifre, L., van den Driessche, G., Schrittwieser, J., Antonoglou, I., Panneershelvam, V., Lanctot, M., Dieleman, S., Grewe, D., Nham, J., Kalchbrenner, N., Sutskever, I., Lillicrap, T., Leach, M., Kavukcuoglu, K., Graepel, T., Hassabis, D.: Mastering the game of Go with deep neural networks and tree search. Nature 529, 484 503 (2016)
10. Solovay, R.M., Strassen, V.: A fast Monte-Carlo test for primality. SIAM J. Comput. 6(1), 84–85 (1977)
11. Solovay, R.M., Strassen, V.: Erratum a fast Monte-Carlo test for primality. SIAM J. Comput. 7(1), 118 (1978)

Integrating Factorization Ranked Features in MCTS: An Experimental Study

Chenjun Xiao[✉] and Martin Müller

Computing Science, University of Alberta, Edmonton, Canada
{chenjun,mmueller}@ualberta.ca

Abstract. Recently, *Factorization Bradley-Terry (FBT)* model is introduced for fast move prediction in the game of Go. It has been shown that FBT outperforms the state-of-the-art fast move prediction system of Latent Factor Ranking (LFR). In this paper, we investigate the problem of integrating feature knowledge learned by FBT model in Monte Carlo Tree Search. We use the open source Go program Fuego as the test platform. Experimental results show that the FBT knowledge is useful in improving the performance of Fuego.

1 Introduction

The idea of Monte Carlo Tree Search (MCTS) [2] is to online construct a search tree of game states evaluated by fast Monte Carlo simulations. However in games with large state space, accurate value estimation by simple simulation cannot be easily guaranteed given limited search time. The inaccurate estimation can mislead the growth of the search tree and can severely limit the strength of the program. Thereby, it is reasonable to incorporate the domain knowledge of the game to serve as heuristic information that benefits the search.

In Computer Go [10] research, knowledge is usually represented by features, such as shape patterns and tactical features. A *move prediction system* applies machine learning techniques to acquire the feature knowledge from professional game records or self played games. Selective search algorithm can then focus on the most promising moves evaluated by such system [1]. For example, [3] proposes Minorization-Maximization (MM) to learn feature knowledge offline and uses it to improve random simulation. [7] considers feature knowledge as a prior to initial statistical values when a new state is added to the search tree. AlphaGo [12] incorporates supervised learned Deep Convolutional Neural Networks (DCNN) as part of in-tree policy for move selection, and further improve the network with reinforcement learning from games of self-play to get a powerful value estimation function. The integrated system becomes the first program to ever beat the world's top Go player.

Recently, [15] introduces *Factorization Bradley-Terry (FBT)* model to learn feature knowledge which became the state-of-the-art fast move prediction algorithm. The major innovation of FBT model is to consider the interaction between different features as part of a probability-based framework, which can be considered as a combination of two leading approaches: MM [3] and LFR [14]. However,

© Springer International Publishing AG 2017
T. Cazenave et al. (Eds.): CGW 2016/GIGA 2016, CCIS 705, pp. 34–43, 2017.
DOI: 10.1007/978-3-319-57969-6_3

it is still not clear whether the feature knowledge learned by this model is useful to improve the strength of the MCTS framework. We investigate this problem in this paper by integrating FBT based knowledge in the open source program Fuego [5].

The remaining of this paper is organized as follows: Sect. 2 describes the idea of FBT model for move prediction in Go. Section 3 discusses how to integrate FBT based knowledge within MCTS framework. Section 4 describes the feature knowledge and move selection scheme in current Fuego and provides the experimental results. Section 5 gives a conclusion of this work and discusses the possible future work.

2   Factorization Bradley-Terry Model for Move Prediction Problem

We briefly describe how FBT model works for move prediction problem in the game of Go. In most popular high-speed move prediction systems, each move is represented as a combination of a group of features. Weights for each feature are learned from expert game records by supervised learning algorithms, and an evaluation function based on the weights is defined to rank moves.

Specifically, let S be the set of possible Go positions, $\Gamma(s)$ be the set of legal moves in a specific position $s \subset S$, and $\mathcal{F}$ be the set of features which are used to describe moves in a given game state. Each move is represented by its set of active features $\mathcal{G} \subseteq \mathcal{F}$. The training set $\mathcal{D}$ consists of cases $\mathcal{D}_j$, with each case representing the possible move choices in one game position s_j, and the expert move is specified as $\mathcal{G}_j^*$.

$$\mathcal{D}_j = \{ \, \mathcal{G}_j^i \mid for \; i = 1, \ldots, |\Gamma(s_j)| \}$$

Most high-speed move prediction systems usually differ from the method of predicting $\mathcal{G}_j^*$ from $\mathcal{D}_j$ as well as the model of the strength of $\mathcal{G}$. In MM [3], the strength of a group $\mathcal{G}$ is approximated by the sum of weights of all features within the group. Prediction of $\mathcal{G}_j^*$ is formulated as a competition among all possible groups. A simple probabilistic model named Generalized Bradley-Terry model [8] defines the probability of each feature group winning a competition. While in another efficient move prediction algorithm called Latent Factor Ranking (LFR) [14], the strength of a group is modelled using a Factorization Machine (FM), which also takes pairwise interactions between features into account besides the sum of all features' weights. Prediction of $\mathcal{G}_j^*$ is simply formulated as a binary classification problem, with $\mathcal{G}_j^*$ in one class and all other groups in the other. LFR outperforms MM in terms of move prediction accuracy. But the evaluation function LFR produces does not provide a probability distribution over all possible moves, which makes it much harder to combine with other kinds of knowledge.

FBT model takes advantage of both MM and LFR: it considers the interaction between features within a group, and produce the evaluation function in a probability-based framework. In FBT, the strength of a group $\mathcal{G} \subseteq \mathcal{F}$ is defined in a same way as in LFR

$$E_{\mathcal{G}} = \sum_{f \in \mathcal{G}} w_f + \frac{1}{2} \sum_{f \in \mathcal{G}} \sum_{g \in \mathcal{G}, g \neq f} \langle v_f, v_g \rangle$$

where $w_f \in \mathbb{R}$ is the (estimated) *strength*, and $v_f \in \mathbb{R}^k$ is the *factorized interaction vector*, of a feature $f \in \mathcal{F}$. The *interaction strength* between two features f and g is modeled as $\langle v_f, v_g \rangle = \sum_{i=1}^{k} v_{f,i} \cdot v_{g,i}$, where k is the pre-defined dimension of the factorization. In Computer Go, setting $k = 5$ and $k = 10$ are most popular [14]. Richer feature sets might require larger k for best performance. With the definition of $E_{\mathcal{G}}$, FBT then applies the Generalized Bradley-Terry model for each test case $\mathcal{D}_j$,

$$P(\mathcal{D}_j) = \frac{exp(E_{\mathcal{G}_j^*})}{\sum_{i=1}^{|\Gamma(s_j)|} exp(E_{\mathcal{G}_j^i})}$$

Suitable parameters in FBT are estimated by maximizing the likelihood of the training data, using a Stochastic Gradient Decent (SGD) algorithm. [15] also provides two techniques to accelerate the training process: an efficient incremental implementation of gradient update, and an unbiased approximate gradient estimator. Details of these two techniques as well as the induction of parameter update formula can be found in [15].

3   Integrating FBT Knowledge in MCTS

As suggested before, a move prediction system can provide useful initial recommendations of which moves are likely to be the best. Selective search with proper exploration scheme, such as MCTS, can further improve upon these recommendations with online simulation information. One favorable property of FBT model is to produce a probability based evaluation. Intuitively, it is a probability distribution of which move is going to be selected by a human expert under a game state. Therefore, it seems very straightforward to incorporate FBT knowledge as part of exploration, since we should explore more on moves which are most favored by human experts.

We apply a variant of PUCT [11] formula which is used in AlphaGo [12] to integrate FBT knowledge in MCTS. The idea of this formula is to explore moves according to a value that is proportional to the predicted probability but decays with repeated visits as in original UCT style [9]. When a new game state s is added to the search tree, we call a pre-trained FBT model to get a prediction $P_{FBT}(s, a)$, which assigns an exploration bonus $E_{FBT}(s, a)$ for each move $a \in \Gamma(s)$. In order to keep sufficient exploration, we set a lower cut threshold λ_{FBT}, where for all $a \in \Gamma(s)$ if $P_{FBT}(s, a) < \lambda_{FBT}$ then simply let

$E_{FBT}(s,a) = \lambda_{FBT}$, otherwise $E_{FBT}(s,a) = P_{FBT}(s,a)$. At state s during in-tree move selection, the algorithm will select the move

$$a = \text{argmax}_{a'}(Q(s,a') + c_{puct}E_{FBT}(s,a')\sqrt{\frac{\lg(N(s))}{1+N(s,a')}}) \qquad (1)$$

where $Q(s,a)$ is the accumulated move value estimated by online simulation, c_{puct} is an exploration constant, $N(s,a)$ is the number of visit time of move a in s, and $N(s) = \sum_i N(s,i)$.

4 Experiments

We use the open source program Fuego [5] as our experiment platform to test if FBT knowledge is helpful for improving MCTS. We first introduce the feature knowledge in current Fuego system, then introduce the training settlement for the FBT model and the setup for the experiment, and finally present the results.

4.1 Feature Knowledge for Move Selection in Fuego

Prior Feature Knowledge. The latest Fuego (SVN version 2017) applies feature knowledge to initialize statistical information when a new state is added to the search tree. A set of features trained with LFR [14] is used where interaction dimension is set at $k = 10$. Since the evaluation LFR produces is a real value indicating the strength of the move without any probability based interpretation, Fuego designed a well-tuned formula to transfer the output value to the prior knowledge for initialization. It adopts a similar method as suggested in [7], where the prior knowledge contains two parts: $N_{prior}(s,a)$ and $Q_{prior}(s,a)$. This indicates that MCTS would perform $N_{prior}(s,a)$ simulations to achieve an estimate of $Q_{prior}(s,a)$ accuracy. Let $V_{LFR}(s,a)$ be the evaluation of move $a \in \Gamma(s)$, $V_{largest}$ and $V_{smallest}$ be the largest and smallest evaluated value respectively. Fuego uses the following formula to assign $N_{prior}(s,a)$ and $Q_{prior}(s,a)$,

$$N_{prior}(s,a) = \begin{cases} \frac{c_{LFR}*|\Gamma(s)|}{SA} * V_{LFR}(s,a) & \text{if } V_{LFR}(s,a) \geq 0 \\ -\frac{c_{LFR}*|\Gamma(s)|}{SA} * V_{LFR}(s,a) & \text{if } V_{LFR}(s,a) < 0 \end{cases} \qquad (2)$$

$$Q_{prior}(s,a) = \begin{cases} 0.5 * (1 + V_{LFR}(s,a)/V_{largest}) & \text{if } V_{LFR}(s,a) \geq 0 \\ 0.5 * (1 - V_{LFR}(s,a)/V_{smallest}) & \text{if } V_{LFR}(s,a) < 0 \end{cases} \qquad (3)$$

where $SA = \sum_i |V_{LFR}(s,i)|$ is the sum of absolute value of each move's evaluation. When a new game state is added to the search tree, Fuego will call the method showed above to initialize the state's statistical information by setting $N(s,a) \leftarrow N_{prior}(s,a)$ and $Q(s,a) \leftarrow Q_{prior(s,a)}$.

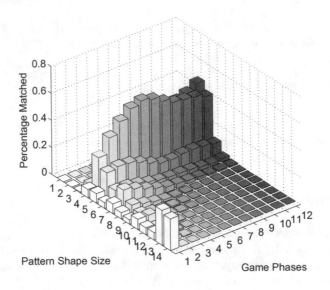

Fig. 1. Distribution of patterns with different size harvested at least 10 times in different game phases.

Greenpeep Knowledge. Another kind of knowledge Fuego also has as part of in-tree move selection policy is called Greenpeep Knowledge. It uses a pre-defined table to get a probability based knowledge $P_g(s, a)$ about each move $a \in \Gamma s$. Then the knowledge is added as a bias for move selection according to the PUCT formula [11]. The reason why Fuego does not use LFR knowledge to replace Greenpeep knowledge might be that LFR cannot produce probability based evaluation. Details can be found in the Fuego source code base [4].

Move Selection in Fuego. In summary, Fuego adopts the following formula to select moves during in-tree search,

$$a = \text{argmax}_{a'}(Q(s, a') - \frac{c_g}{\sqrt{P_g(s, a')}} \times \sqrt{\frac{N(s, a')}{N(s, a') + 5}}) \tag{4}$$

where c_g is a parameter controlling the scale of the Greenpeep knowledge. $Q(s, a')$ is initialized according to Eqs. (2) and (3), and further improved with Monte Carlo simulation and Rapid Action Value Estimation (RAVE). Note that formula (4) does not have the UCB style exploration term, since the exploration constant is set to zero in Fuego. The only exploration comes from RAVE. Comparing formula (4) with (1), we could consider the FBT knowledge $P_{FBT}(s, a)$ as a replacement of the Greenpeep knowledge $P_g(s, a)$, but with a different way to be added as a bias and different decay function.

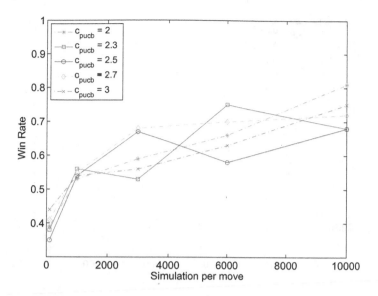

Fig. 2. Experimental results: FBT-FuegoNoLFR vs FuegoNoLFR.

4.2 Training Settlement for FBT Model

We train a FBT model with interaction dimension $k = 5$ using 10000 master games download from the public domain at https://badukmovies.com/pro_games. The prediction accuracy of this model is 38.26%. The parameters of the training algorithm including learning rate and regularization parameters are set at the same as described in [15]. We also apply the same stopping criteria that the training process is stopped and the best performing weight set is returned if the prediction accuracy on a validation set does not increase for three iterations.

The simple features used in this work are listed below. Most features are the same as suggested in [15]. We only use large pattern as the shape pattern for this experiment. All patterns are harvested as in [13,14]. Figure 1 shows the distribution of harvested largest matches for the different pattern sizes in each game phase. The implementation of the tactical features is part of the Fuego program [5], details can be found in the Fuego code base [4]. Note that current Fuego includes the same set of tactical features. But it uses small shape patterns instead of large patterns for feature knowledge evaluation.

- **Pass**
- **Capture, Extension, Self-atari, Atari** Tactical features similar to [3].
- **Line and Position (edge distance perpendicular to Line)** ranges from 1 to 10.
- **Distance to previous move** feature values are $2, \ldots, 16, \geq 17$. The distance is measured by $d(\delta x, \delta y) = |\delta x| + |\delta y| + max\{|\delta x|, |\delta y|\}$.
- **Distance to second-last move** uses the same metric as previous move. The distance can be 0.

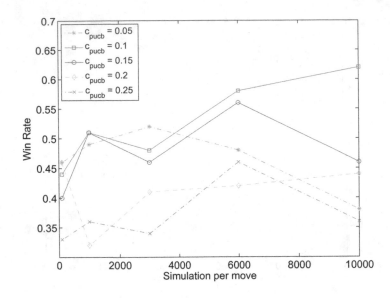

Fig. 3. Experimental results: FBT-Fuego vs Fuego.

- **Fuego Playout Policy.** These features correspond to the rules in the playout policy used in Fuego. Most are simple tactics related to stones with a low number of liberties and 3×3 patterns.
- **Side Extension.** The distance to the closest stones along the sides of the board.
- **Corner Opening Move.** Standard opening moves.
- **CFG Distance.** The distance when contracting all stones in a block to a single node in a graph [6].
- **Shape Patterns.** Circular patterns with sizes from 2 to 14. All shape patterns are invariant to rotation, translation and mirroring.

4.3   Setup

Experiments are performed on a 2.4 GHz Intel Xeon CPU with 64 GB memory. We use the latest Fuego (SVN revision 2017) in the experiment. We call Fuego without LFR prior knowledge as FuegoNoLFR, Fuego applying formula (1) to select moves in-tree as FBT-Fuego, and Fuego without LFR but using formula (1) as FBT-FuegoNoLFR. The lower cut threshold for FBT knowledge is set to $\lambda_{FBT} = 0.001$. All other parameters are default as in the original settings of Fuego.

4.4   Experimental Results

We first compare FBT-FuegoNoLFR with FuegoNoLFR. This experiment is designed to show the strength of FBT knowledge without any influence from

Table 1. Running time comparison (specified in seconds) with different simulations for per move.

Program name	100	1000	3000	6000	10000
FBT-FuegoNoLFR	11.8	192.4	704.1	1014.5	1394.9
FuegoNoLFR	5.1	55.7	148.7	225.2	354.4
FBT-Fuego	23.4	241.1	734.1	912.6	1417.5
Fuego	10.8	168.3	564.2	778.6	1161.2

other kinds of knowledge. We test the performance of FBT-FuegoNoLFR against FuegoNoLFR with different exploration constants c_{puct}. After initial experiments, the range explored was $c_{puct} \in \{2, 2.3, 2.5, 2.7, 3\}$. In order to investigate if the FBT knowledge is scaling with the number of simulations per move, N_{sim} was tested by setting $N_{sim} \in \{100, 1000, 3000, 6000, 10000\}$. Figure 2 shows the win rate of FBT-FuegoNoLFR against FuegoNoLFR. All data points are averaged over 1000 games. The results show that adding FBT knowledge can dramatically improve the performance of Fuego over the baseline without feature knowledge as prior. FBT-FuegoNoLFR scales well with more simulations per move. With $c_{pucb} = 2$ and 10000 simulations per move FBT-FuegoNoLFR can beat FuegoNoLFR with 81% winning rate.

We then compare FBT-Fuego with full Fuego, in order to investigate if the FBT knowledge is comparable with current feature knowledge in Fuego and able to improve the performance in general. In this case, c_{puct} is tuned over a different range, $c_{puct} \in \{0.05, 0.1, 0.15, 0.2, 0.25\}$. $N_{sim} \in \{100, 1000, 3000, 6000, 10000\}$, and all data points are averaged over 1000 games as before. Results are presented in Fig. 3. FBT-Fuego has worse performance in most settings of c_{puct}. But it can be made to work after careful tuning. As suggested in Fig. 3, under the setting where $c_{pucb} = 0.1$, FBT-Fuego scales well with the number of simulations per move, and achieves 62% winning rate against Fuego with 10000 simulations per move. One possible reason is that the FBT knowledge is not quite comparable with the LFR knowledge. The moves these two methods favour might be different in some situations, which makes it very hard to tune a well-tuned system when adding another knowledge term.

Finally, we show the running time of our methods with different simulations per move in Table 1. FBT-FuegoNoLFR spends much more time than FuegoNoLFR, since FuegoNoLFR only uses Greenpeep knowledge for exploration and thus does not need to compute any feature knowledge. FBT-FuegoNoLFR spends a little less time than FBT-Fuego, since it does not use feature knowledge to initialize prior knowledge. The speed of FBT-Fuego is a little worse than Fuego. The time difference is spent on computing large patterns, while Fuego only uses small shape patterns.

5   Conclusion and Future Work

In this paper, we introduce how to integrate the state-of-the-art fast move prediction algorithm FBT in MCTS. We use the open source program Fuego as our test platform. Experimental results show that FBT knowledge is useful to improve the performance of Fuego, without too much sacrifice in efficiency.

Future work includes: 1. try to discover a method to transform FBT knowledge as prior knowledge for initialization. 2. try to apply the FBT knowledge for improving fast roll-out policy, which has been shown as a very important part in the state-of-the-art Go program AlphaGo [12].

References

1. Browne, C., Powley, E., Whitehouse, D., Lucas, S., Cowling, P., Rohlfshagen, P., Tavener, S., Perez, D., Samothrakis, S., Colton, S.: A survey of Monte Carlo tree search methods. IEEE Trans. Comput. Intellig. AI Games **4**(1), 1–43 (2012)
2. Coulom, R.: Efficient selectivity and backup operators in Monte-Carlo tree search. In: Herik, H.J., Ciancarini, P., Donkers, H.H.L.M.J. (eds.) CG 2006. LNCS, vol. 4630, pp. 72–83. Springer, Heidelberg (2007). doi:10.1007/978-3-540-75538-8_7
3. Coulom, R.: Computing "Elo ratings" of move patterns in the game of Go. ICGA J. **30**(4), 198–208 (2007)
4. Enzenberger, M., Müller, M.: Fuego (2008–2015). http://fuego.sourceforge.net
5. Enzenberger, M., Müller, M., Arneson, B., Segal, R.: Fuego - an open-source framework for board games and Go engine based on Monte Carlo tree search. IEEE Trans. Comput. Intell. AI Games **2**(4), 259–270 (2010)
6. Friedenbach, K.J.: Abstraction hierarchies: a model of perception and cognition in the game of Go. Ph.D. thesis, University of California, Santa Cruz (1980)
7. Gelly, S., Silver, D.: Combining online and offline knowledge in UCT. In: Proceedings of the 24th International Conference on Machine Learning (ICML 2007), pp. 273–280. ACM (2007)
8. Hunter, D.R.: MM algorithms for generalized Bradley-Terry models. Ann. Stat. **32**(1), 384–406 (2004)
9. Kocsis, L., Szepesvári, C.: Bandit based Monte-Carlo planning. In: Fürnkranz, J., Scheffer, T., Spiliopoulou, M. (eds.) ECML 2006. LNCS (LNAI), vol. 4212, pp. 282–293. Springer, Heidelberg (2006). doi:10.1007/11871842_29
10. Müller, M.: Computer Go. Artif. Intell. **134**(1–2), 145–179 (2002)
11. Rosin, C.: Multi-armed bandits with episode context. Ann. Math. Artif. Intell. **61**(3), 203–230 (2011)
12. Silver, D., Huang, A., Maddison, C.J., Guez, A., Sifre, L., van den Driessche, G., Schrittwieser, J., Antonoglou, I., Panneershelvam, V., Lanctot, M., Dieleman, S., Grewe, D., Nham, J., Kalchbrenner, N., Sutskever, I., Lillicrap, T., Leach, M., Kavukcuoglu, K., Graepel, T., Hassabis, D.: Mastering the game of Go with deep neural networks and tree search. Nature **529**(7587), 484–489 (2016)
13. Stern, D., Herbrich, R., Graepel, T.: Bayesian pattern ranking for move prediction in the game of Go. In: Proceedings of the 23rd International Conference on Machine Learning, pp. 873–880. ACM (2006)

14. Wistuba, M., Schmidt-Thieme, L.: Move prediction in Go – modelling feature interactions using latent factors. In: Timm, I.J., Thimm, M. (eds.) KI 2013. LNCS (LNAI), vol. 8077, pp. 260–271. Springer, Heidelberg (2013). doi:10.1007/978-3-642-40942-4_23
15. Xiao, C., Müller, M.: Factorization ranking model for move prediction in the game of Go. In: AAAI, pp. 1359–1365 (2016)

Nested Rollout Policy Adaptation
with Selective Policies

Tristan Cazenave[✉]

PSL-Université Paris-Dauphine, LAMSADE CNRS UMR 7243, Paris, France
cazenave@lamsade.dauphine.fr

Abstract. Monte Carlo Tree Search (MCTS) is a general search algorithm that has improved the state of the art for multiple games and optimization problems. Nested Rollout Policy Adaptation (NRPA) is an MCTS variant that has found record-breaking solutions for puzzles and optimization problems. It learns a playout policy online that dynamically adapts the playouts to the problem at hand. We propose to enhance NRPA using more selectivity in the playouts. The idea is applied to three different problems: Bus regulation, SameGame and Weak Schur numbers. We improve on standard NRPA for all three problems.

1 Introduction

Monte Carlo Tree Search (MCTS) is a state-of-the-art search algorithm that has greatly improved the level of play in games such as Go [12,13] and Hex [22]. The principle underlying MCTS is to play random games and to use the statistics on the moves played during the games so as to find the best moves [25].

MCTS can also be applied to problems other than games [6]. Examples of non-games applications are Security, Mixed Integer Programming, Traveling Salesman Problem, Physics Simulations, Function Approximation, Constraint Problems, Mathematical Expressions, Planning and Scheduling.

Some MCTS algorithms have been tailored to puzzles and optimization problems. For example Nested Monte Carlo Search (NMCS) [7] gives good results for multiple optimization problems. NRPA is an improvement of NMCS that learns a playout policy online [28].

In this paper we improve NRPA adding selectivity in the playouts. We propose to modify the standard playout policy used by NRPA in order to avoid bad moves during playouts.

The paper is organized in three remaining sections. Section 2 presents related works, Sect. 3 details selective policies for different problems and Sect. 4 gives experimental results.

2 Related Work

NMCS is an algorithm that improves Monte Carlo search with Monte Carlo search. It has different levels of nested playouts and an important feature of the

© Springer International Publishing AG 2017
T. Cazenave et al. (Eds.): CGW 2016/GIGA 2016, CCIS 705, pp. 44–56, 2017.
DOI: 10.1007/978-3-319-57969-6_4

algorithm is that it records the best sequence of moves at each search level. The algorithm was initially applied to puzzles such as Morpion Solitaire, SameGame and Sudoku.

A further application of NMCS is the Snake-In-The-Box problem [23]. NMCS has beaten world records for this problem that has application in coding theory. The goal of the problem is to find the longest possible path in a high dimensional hypercube so that nodes in the path never have more than two neighboring nodes also in the path.

Bruno Bouzy improved NMCS by using playout policies. For example for the Pancake problem [4] he uses domain specific playout policies so as to beat world records with an improved NMCS. Another variation on NMCS is Monte-Carlo Fork Search [2] that branches deep in the playouts. It was successfully applied to complex cooperative pathfinding problems.

The Weak Schur problem [19] is a problem where informed playout policies can give much better results than standard policies when used with NMCS [3]. Policy learning has been successfully used for the Traveling Salesman with Time Windows (TSPTW) in combination with NMCS [27].

An effective combination of nested levels of search and of policy learning has been proposed with the NRPA algorithm [28]. NRPA holds world records for Morpion Solitaire and crosswords puzzles. NRPA is given in Algorithm 3. The principle is to learn weights for the possible actions so as to bias the playouts. The playout algorithm is given in Algorithm 1. It performs Gibbs sampling, choosing the actions with a probability proportional to the exponential of their weights. The weights of the actions are updated at each step of the algorithm so as to favor moves of the best sequence found so far at each level. The principle of the adaptation is to add 1.0 to the action of the best sequence and to decrease the weight of the other possible actions by an amount proportional to the exponential of their weight. The adaptation algorithm is given in Algorithm 2.

Playout policy adaptation has also been used for games such as Go [20] or various other games with success [8].

NRPA works by biasing the probability of an action by looking up a weight associated to the action. An alternative is to make the bias a function of the current state and the proposed action [26]. An improvement of standard NRPA is to combine it with beam search yielding Beam NRPA [11].

Stefan Edelkamp and co-workers have applied the NRPA algorithm to multiple problems. They have optimized the algorithm for the TSPTW problem [10, 14]. Other applications deal with 3D packing with object orientation [16], the physical traveling salesman problem [17], the multiple sequence alignment problem [18], logistics [15] or cryptography [21].

3   Selective Policies

The principle underlying selective policies is to modify the legal moves so that moves that are unlikely to be good are pruned during playouts.

Algorithm 1. The playout algorithm

---

playout $(state, policy)$
$sequence \leftarrow []$
while true **do**
  if $state$ is terminal **then**
    return (score $(state)$, $sequence$)
  end if
  $z \leftarrow 0.0$
  for m in possible moves for $state$ **do**
    $z \leftarrow z + \exp(k \times policy\,[\text{code}(m)])$
  end for
  choose a $move$ with probability proportional to $\frac{exp(k \times policy[code(move)])}{z}$
  $state \leftarrow$ play $(state, move)$
  $sequence \leftarrow sequence + move$
end while

Algorithm 2. The adapt algorithm

---

adapt $(policy, sequence)$
$polp \leftarrow policy$
$state \leftarrow root$
for $move$ in $sequence$ **do**
  $polp\,[\text{code}(move)] \leftarrow polp\,[\text{code}(move)] + \alpha$
  $z \leftarrow 0.0$
  for m in possible moves for $state$ **do**
    $z \leftarrow z + \exp(policy\,[\text{code}(m)])$
  end for
  for m in possible moves for $state$ **do**
    $polp\,[\text{code}(m)] \leftarrow polp\,[\text{code}(m)] - \alpha * \frac{exp(policy[code(m)])}{z}$
  end for
  $state \leftarrow$ play $(state, move)$
end for
$policy \leftarrow polp$

Algorithm 3. The NRPA algorithm

NRPA $(level, policy)$
if level $== 0$ **then**
  return playout (root, $policy$)
end if
$bestScore \leftarrow -\infty$
for N iterations **do**
  (result,new) $\leftarrow$ NRPA$(level - 1, policy)$
  if result $\geq$ bestScore **then**
    bestScore $\leftarrow$ result
    seq $\leftarrow$ new
  end if
  policy $\leftarrow$ adapt (policy, seq)
end for
return (bestScore, seq)

This can be done differently for each application of the algorithm. In this section we describe the move pruning for three problems: the bus regulation problem, SameGame and the Weak Schur problem.

3.1   Bus Regulation

In the bus regulation problem [9] the bus regulator knows the location of all the buses of a bus line. At each stop he can decide to make a bus wait before continuing his route. Waiting at a stop can reduce the overall passengers waiting time. The score of a simulation is the sum of all the passengers waiting time. Optimizing a problem is finding a set of bus stopping times that minimizes the score of the simulation. It is possible to use rules to decide the bus waiting time given the number of stops before the next bus. Monte Carlo bus regulation with NMCS has been shown to improve on rule-based regulation.

In this paper we use NRPA to choose the bus waiting times. We compare the standard policy that can choose a waiting time between 1 and 5 min to a selective policy that always chooses a waiting time of 1 if there are fewer than δ stops before the next bus.

An important detail of the NRPA algorithm is the way moves are coded. A move code for the bus regulation problem takes into account the bus stop, the time of arrival to the bus stop and the number of minutes to wait before leaving the stop.

Algorithm 4 gives the rule used to compute the legal moves for the bus regulation problem.

Algorithm 4. Legal moves with a selective policy for the bus regulation problem.

legalMoves (*moves*)
moves ← [1 min]
if next bus is at strictly less than δ stops **then**
    return *moves*
end if
for i in 2,max waiting time **do**
    add (i minutes) to *moves*
end for
return *moves*

3.2   SameGame

SameGame is a puzzle composed of a rectangular grid containing cells of different colors. A move removes connected cells of the same color. The cells of other colors fall to fill the void created by a move. At least two cells have to be removed for a move to be legal. The score of a move is the square of the number of removed cells minus two. A bonus of one thousand is credited for completely clearing the board.

MCTS has been quite successful for SameGame. SP-MCTS [29,30], NMCS [7] and Nested MCTS [1] have reached great scores at SameGame. For all algorithms an effective improvement on random playouts is to use the tabu color strategy. As it is often beneficial to remove all the cells of the most frequent color in one move, the tabu color strategy avoids the moves of the most frequent color until all of its cells form only one group.

We propose to apply NRPA to SameGame and to improve on standard NRPA using selective policies.

There are many possible different moves at SameGame. So many moves that it is not possible to code them with a simple function without exceeding storage capacities. The way we deal with this problem is by using Zobrist hashing [31]. Zobrist hashing is popular in computer games such as Go and Chess [5]. It uses a 64 bits random integer for each possible color of each cell of the board. The code for a move is the XOR of the random numbers associated to the cells of the move. A transposition table is used to store the codes and their associated weights. The index of a move in the transposition table is its 16 lower bits. For each entry of the transposition table, a list of move codes and weights is stored.

It has been previously shown that in SameGame it is possible to improve simulation policies by allowing more randomness in the endgame [24].

What we do is that we use a modified version of the tabu color strategy. We allow moves of size two of the tabu color when the number of moves already played is greater than a threshold with value t. Algorithm 5 gives the function used to compute the legal moves for SameGame.

Algorithm 5. Legal moves with a selective policy for SameGame.

---

legalMoves (*moves*, tabuColor)
if only one move of the tabu color **then**
    tabuColor = noColor
end if
for m in possible moves **do**
    if color (m) == tabuColor **then**
        if nbCells (m) == 2 and nb moves played > t **then**
            add m to *moves*
        end if
    else
        add m to *moves*
    end if
end for
if *moves* is empty **then**
    for m in possible moves **do**
        add m to *moves*
    end for
end if

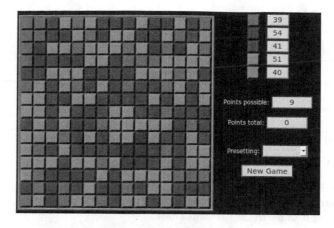

Fig. 1. Example of the initial state of a SameGame problem (Color figure online)

Figure 1 gives an example of a starting board at SameGame. We can see on the side the number of cells for each color. In this example the tabu color is green since it has 54 cells.

3.3 Weak Schur Numbers

The Weak Schur problem is to find a partition of consecutive numbers that contains as many consecutive numbers as possible, where a partition must not contain a number that is the sum of two previous numbers in the same partition.

The last number that was added to the partition before the next number could not be placed is the score of a partition. The goal is to find partitions with high scores.

The current records for the Weak Schur problem are given in Table 1. The records for 7 and 8 are held by Bruno Bouzy using NMCS [3].

One of the best partitions of size three is for example:

```
1 2 4 8 11 22
3 5 6 7 19 21 23
9 10 12 13 14 15 16 17 18 20
```

When possible, it is often a good move to put the next number in the same partition as the previous number. The selective policy for SameGame follows this heuristic. The algorithm for the legal moves is given in Algorithm 6. If it is legal to put the next number n in the same partition as the previous number then it is the only legal move considered. Otherwise all legal moves are considered.

The code of a move for the Weak Schur problem takes as input the partition of the move, the integer to assign and the previous number in the partition.

Table 1. Current records for the weak scwur problem

K	1	2	3	4
WS(K)	$= 2$	$= 8$	$= 23$	$= 66$
K	5	6	7	8
WS(K)	≥ 196	≥ 582	≥ 1736	≥ 5105

Algorithm 6. Legal moves with a selective policy for the Weak Schur problem.

legalMoves (*moves*, *n*)
moves ← []
for *i* in 0,nbPartitions **do**
   if previous number in partition i == n - 1 **then**
      if playing n in partition i is legal **then**
         add (i, n) to *moves*
      end if
   end if
end for
if *moves* == [] **then**
   for *i* in 0,nbPartitions **do**
      if playing n in partition i is legal **then**
         add (i, n) to *moves*
      end if
   end for
end if

---

4   Experimental Results

In order to evaluate a policy we run 200 times the NRPA algorithm with this policy. The scores are recorded starting at 0.01 s and for every power of two multiplied by 0.01. The algorithm is stopped after 163.84 s. We chose to record the scores this way in order to see the average improvement in score each time the search time is doubled. It has no influence on the NRPA algorithm.

4.1   Bus Regulation

Table 2 gives the evolution with time of the best score of the standard NRPA algorithm in the No δ column and compares it to the evolution of the best score using rules with different δ for the legal moves. We can see that using $\delta = 3$ always gives better results than the No δ policy. For small times the $\delta = 4$ policy is much better than the other policies. For the longest search time (163.84 s), the playout rule that uses $\delta = 3$ has a score of 1,610 which is better than the playout policy without rules that has a score of 1,632.

Table 2. Evaluation of selective policies for the bus regulation problem

Time	No δ	$\delta = 1$	$\delta = 2$	$\delta = 3$	$\delta = 4$
0.01	2,620	2,441	2,344	2,147	1,929
0.02	2,441	2,292	2,173	2,049	1,866
0.04	2,329	2,224	2,098	2,000	1,828
0.08	2,242	2,178	2,045	1,959	1,791
0.16	2,157	2,135	2,011	1,925	1,764
0.32	2,107	2,108	1,986	1,903	1,736
0.64	2,046	2,074	1,959	1,868	1,713
1.28	1,974	2,013	1,917	1,811	1,694
2.56	1,892	1,926	1,869	1,754	1,679
5.12	1,802	1,832	1,822	1,703	1,667
10.24	1,737	1,757	1,769	1,660	1,658
20.48	1,698	1,712	1,729	1,640	1,651
40.96	1,682	1,695	1,699	1,629	1,644
81.92	1,660	1,674	1,661	1,617	1,637
163.84	1,632	1,642	1,629	**1,610**	1,635

4.2   SameGame

We performed two experiments for SameGame. The first experiment tests different playout strategies for the first problem of the test set. NRPA is run 200 times for each strategy and the evolution of the mean score with time is recorded.

The second experiment runs a level 4 search on the standard test set and the results are compared to the state of the art.

Table 3 gives the evolution of the mean score for problem one of the standard test set. We can observe that the tabu strategy is a large improvement over the standard policy (2,484.18 instead of 2,011.25). Allowing moves of the tabu color of size two when the playout length is greater than 10 gives even better results for long time settings even if it is worse for short time settings. The tabu policy is equivalent to the selective policy with $t = \infty$. For short time settings the tabu policy is the best one. However, when more time is given to the algorithm it discovers ways of using the increased freedom of moves contained in the selective policy with $t > 10$ and eventually reaches a better score of 2,636.22 instead of 2,484.18 for the tabu policy.

Table 4 gives the best scores obtained with different algorithms for SameGame. The website js-games.de maintains the best scores obtained by its internet users. We can see that these scores are higher than the one obtained with Monte Carlo search. Little is known about the holders of these records. However we could exchange emails with a record holder who told us he is using beam search with a complex domain specific evaluation function to play SameGame.

Table 3. Evaluation of selective policies for SameGame

Time	No tabu	tabu	$t > 0$	$t > 10$
0.01	155.83	352.19	260.37	257.59
0.02	251.28	707.56	487.27	505.05
0.04	340.18	927.63	666.91	677.57
0.08	404.27	1,080.64	810.29	822.44
0.16	466.15	1,252.14	924.41	939.30
0.32	545.78	1,375.78	1,043.97	1,058.54
0.64	647.63	1,524.37	1,185.77	1,203.91
1.28	807.20	1,648.16	1,354.69	1,356.81
2.56	1,012.42	1,746.74	1,508.10	1,497.90
5.12	1,184.77	1,819.43	1,616.44	1,605.86
10.24	1,286.25	1,886.48	1,737.35	1,712.17
20.48	1,425.55	1,983.42	1,859.12	1,879.10
40.96	1,579.67	2,115.80	2,078.30	2,100.47
81.92	1,781.40	2,319.44	2,329.73	2,384.24
163.84	2,011.25	2,484.18	2,539.75	**2,636.22**

Table 4. Best scores for SameGame

Position	NMCS	SP-MCTS	Selective NRPA	js-games.de
1	3,121	2,919	3,179	3,413
2	3,813	3,797	3,985	4,023
3	3,085	3,243	3,635	4,019
4	3,697	3,687	3,913	4,215
5	4,055	4,067	4,309	4,379
6	4,459	4,269	4,809	4,869
7	2,949	2,949	2,651	3,435
8	3,999	4,043	3,879	4,771
9	4,695	4,769	4,807	5,041
10	3,223	3,245	2,831	3,937
11	3,147	3,259	3,317	3,783
12	3,201	3,245	3,315	3,921
13	3,197	3,211	3,399	3,821
14	2,799	2,937	3,097	3,263
15	3,677	3,343	3,559	4,161
16	4,979	5,117	5,025	5,517
17	4,919	4,959	5,043	5,227
18	5,201	5,151	5,407	5,503
19	4,883	4,803	5,065	5,343
20	4,835	4,999	4,805	5,217
Total	77,934	78,012	80,030	**87,858**

We can also observe that NRPA with a selective policy has better scores than NMCS and SP-MCTS since the total of its scores is 80,030 for a level 4 search. It is approximately 2,000 points better than previous MCTS algorithms.

Table 5. Evaluation of selective policies for the Weak Schur problem

Time	ws(6)	ws-rule(6)	ws(7)	ws-rule(7)
0.01	81	300	111	652
0.02	110	376	150	825
0.04	117	398	160	901
0.08	123	419	168	950
0.16	129	435	177	1,001
0.32	137	448	186	1,050
0.64	147	460	197	1,100
1.28	154	465	216	1,150
2.56	164	468	236	1,184
5.12	174	479	252	1,203
10.24	186	489	267	1,220
20.48	197	498	284	1,258
40.96	215	503	303	1,297
81.92	232	505	337	1,332
163.84	239	**506**	384	**1,356**

Table 6. Evaluation of selective policies for the Weak Schur problem

Time	ws(8)	ws-rule(8)	ws(9)	ws-rule(9)
0.01	151	1,382	199	2,847
0.02	193	1,707	246	3,342
0.04	207	1,898	263	3,717
0.08	218	2,055	273	4,125
0.16	227	2,162	286	4,465
0.32	236	2,297	293	4,757
0.64	245	2,423	303	5,044
1.28	263	2,574	314	5,357
2.56	288	2,717	331	5,679
5.12	316	2,852	362	6,065
10.24	335	2,958	384	6,458
20.48	351	3,010	403	6,805
40.96	371	3,096	422	7,117
81.92	394	3,213	444	7,311
163.84	440	**3,318**	473	**7,538**

4.3 Weak Schur Numbers

Tables 5 and 6 give the evolution with time of the best score of the standard NRPA algorithm and of the rule-based selective NRPA algorithm. The most striking example of the usefulness of a selective policy is for 9 partitions in Table 6. The standard policy reaches 473 in 163.84 s when the selective policy reaches 7,538 for the same running time.

5 Conclusion

We have applied selective policies to three quite different problems. For each of these problems selective policies improve NRPA. We only used simple policy improvements, better performance could be obtained refining the proposed policies.

For all three problems, simple and effective rules could be found that avoid bad moves in playouts. In some other problems such as Morpion Solitaire [7,28] for example such rules could be more difficult to find. Also, even if rules generally improve playouts they can make NRPA blind to some moves that are good in specific cases and prevent it from finding the best sequence.

For future work we intend to improve the selective policies and to apply the principle to other difficult problems.

References

1. Baier, H., Winands, M.H.M.: Nested Monte-Carlo tree search for online planning in large MDPs. In: ECAI 2012–20th European Conference on Artificial Intelligence, pp. 109–114. IOS press (2012)
2. Bouzy, B.: Monte-Carlo fork search for cooperative path-finding. In: Cazenave, T., Winands, M.H.M., Iida, H. (eds.) CGW 2013. CCIS, vol. 408, pp. 1–15. Springer, Cham (2014). doi:10.1007/978-3-319-05428-5_1
3. Bouzy, B.: An abstract procedure to compute Weak Schur number lower bounds. Technical report 2, LIPADE, Université Paris Descartes (2015)
4. Bouzy, B.: An experimental investigation on the pancake problem. In: Cazenave, T., Winands, M.H.M., Edelkamp, S., Schiffel, S., Thielscher, M., Togelius, J. (eds.) CGW/GIGA -2015. CCIS, vol. 614, pp. 30–43. Springer, Cham (2016). doi:10.1007/978-3-319-39402-2_3
5. Breuker, D.M.: Memory versus search in games. Ph.D. thesis, Universiteit Maastricht, Maastricht, The Netherlands (1998)
6. Browne, C., Powley, E., Whitehouse, D., Lucas, S., Cowling, P., Rohlfshagen, P., Tavener, S., Perez, D., Samothrakis, S., Colton, S.: A survey of Monte Carlo tree search methods. IEEE Trans. Comput. Intell. AI Games **4**(1), 1–43 (2012)
7. Cazenave, T.: Nested Monte-Carlo search. In: Boutilier, C. (ed.) IJCAI, pp. 456–461 (2009)
8. Cazenave, T.: Playout policy adaptation with move features. Theor. Comput. Sci. **644**, 43–52 (2016)
9. Cazenave, T., Balbo, F., Pinson, S.: Monte-Carlo bus regulation. In: ITSC, pp. 340–345. St. Louis (2009)

10. Cazenave, T., Teytaud, F.: Application of the nested rollout policy adaptation algorithm to the traveling salesman problem with time windows. In: Hamadi, Y., Schoenauer, M. (eds.) LION 2012. LNCS, pp. 42–54. Springer, Heidelberg (2012). doi:10.1007/978-3-642-34413-8_4

11. Cazenave, T., Teytaud, F.: Beam nested rollout policy adaptation. In: Computer Games Workshop, CGW 2012, ECAI 2012, pp. 1–12 (2012)

12. Coulom, R.: Computing Elo ratings of move patterns in the game of Go. ICGA J. **30**(4), 198–208 (2007)

13. Coulom, R.: Efficient selectivity and backup operators in Monte-Carlo tree search. In: Herik, H.J., Ciancarini, P., Donkers, H.H.L.M.J. (eds.) CG 2006. LNCS, vol. 4630, pp. 72–83. Springer, Heidelberg (2007). doi:10.1007/978-3-540-75538-8_7

14. Edelkamp, S., Gath, M., Cazenave, T., Teytaud, F.: Algorithm and knowledge engineering for the TSPTW problem. In: 2013 IEEE Symposium on Computational Intelligence in Scheduling (SCIS), pp. 44–51. IEEE (2013)

15. Edelkamp, S., Gath, M., Greulich, C., Humann, M., Herzog, O., Lawo, M.: Monte-Carlo tree search for logistics. In: Clausen, U., Friedrich, H., Thaller, C., Geiger, C. (eds.) Commercial Transport. LNL, pp. 427–440. Springer, Cham (2016). doi:10.1007/978-3-319-21266-1_28

16. Edelkamp, S., Gath, M., Rohde, M.: Monte-Carlo tree search for 3D packing with object orientation. In: Lutz, C., Thielscher, M. (eds.) KI 2014. LNCS (LNAI), vol. 8736, pp. 285–296. Springer, Cham (2014). doi:10.1007/978-3-319-11206-0_28

17. Edelkamp, S., Greulich, C.: Solving physical traveling salesman problems with policy adaptation. In: 2014 IEEE Conference on Computational Intelligence and Games (CIG), pp. 1–8. IEEE (2014)

18. Edelkamp, S., Tang, Z.: Monte-carlo tree search for the multiple sequence alignment problem. In: Eighth Annual Symposium on Combinatorial Search (2015)

19. Eliahou, S., Fonlupt, C., Fromentin, J., Marion-Poty, V., Robilliard, D., Teytaud, F.: Investigating Monte-Carlo methods on the Weak Schur problem. In: Middendorf, M., Blum, C. (eds.) EvoCOP 2013. LNCS, vol. 7832, pp. 191–201. Springer, Heidelberg (2013). doi:10.1007/978-3-642-37198-1_17

20. Graf, T., Platzner, M.: Adaptive playouts in Monte-Carlo tree search with policy-gradient reinforcement learning. In: Plaat, A., Herik, J., Kosters, W. (eds.) ACG 2015. LNCS, vol. 9525, pp. 1–11. Springer, Cham (2015). doi:10.1007/978-3-319-27992-3_1

21. Hauer, B., Hayward, R., Kondrak, G.: Solving substitution ciphers with combined language models. In: COLING 2014, 25th International Conference on Computational Linguistics, Proceedings of the Conference: Technical Papers, August 23–29, 2014, Dublin, Ireland, pp. 2314–2325 (2014)

22. Huang, S., Arneson, B., Hayward, R.B., Müller, M., Pawlewicz, J.: MoHex 2.0: a pattern-based MCTS hex player. In: Computers and Games - 8th International Conference, CG 2013, Yokohama, Japan, August 13–15, 2013, Revised Selected Papers, pp. 60–71 (2014)

23. Kinny, D.: A new approach to the snake-in-the-box problem. In: ECAI, vol. 242, pp. 462–467 (2012)

24. Klein, S.: Attacking SameGame using Monte-Carlo tree search: using randomness as guidance in puzzles. Master's thesis, KTH Royal Institute of Technology, Stockholm, Sweden (2015)

25. Kocsis, L., Szepesvári, C.: Bandit based Monte-Carlo planning. In: Fürnkranz, J., Scheffer, T., Spiliopoulou, M. (eds.) ECML 2006. LNCS (LNAI), vol. 4212, pp. 282–293. Springer, Heidelberg (2006). doi:10.1007/11871842_29

26. Lucas, S.M., Samothrakis, S., Pérez, D.: Fast evolutionary adaptation for Monte Carlo tree search. In: Esparcia-Alcázar, A.I., Mora, A.M. (eds.) EvoApplications 2014. LNCS, vol. 8602, pp. 349–360. Springer, Heidelberg (2014). doi:10.1007/978-3-662-45523-4_29

27. Rimmel, A., Teytaud, F., Cazenave, T.: Optimization of the nested Monte-Carlo algorithm on the traveling salesman problem with time windows. In: Chio, C., et al. (eds.) EvoApplications 2011. LNCS, vol. 6625, pp. 501–510. Springer, Heidelberg (2011). doi:10.1007/978-3-642-20520-0_51

28. Rosin, C.D.: Nested rollout policy adaptation for Monte Carlo tree search. In: IJCAI, pp. 649–654 (2011)

29. Schadd, M.P.D., Winands, M.H.M., Tak, M.J.W., Uiterwijk, J.W.H.M.: Single-player Monte-Carlo tree search for SameGame. Knowl. Based Syst. **34**, 3–11 (2012)

30. Schadd, M.P.D., Winands, M.H.M., Herik, H.J., Chaslot, G.M.J.-B., Uiterwijk, J.W.H.M.: Single-player Monte-Carlo tree search. In: Herik, H.J., Xu, X., Ma, Z., Winands, M.H.M. (eds.) CG 2008. LNCS, vol. 5131, pp. 1–12. Springer, Heidelberg (2008). doi:10.1007/978-3-540-87608-3_1

31. Zobrist, A.L.: A new hashing method with application for game playing. ICCA J. **13**(2), 69–73 (1990)

A Rollout-Based Search Algorithm Unifying MCTS and Alpha-Beta

Hendrik Baier[✉]

Advanced Concepts Team European Space Agency, Noordwijk, The Netherlands
hendrik.baier@esa.int

Abstract. *Monte Carlo Tree Search* (MCTS) has been found to be a weaker player than minimax in some tactical domains, partly due to its highly selective focus only on the most promising moves. In order to combine the strategic strength of MCTS and the tactical strength of minimax, *MCTS-minimax hybrids* have been introduced in prior work, embedding shallow minimax searches into the MCTS framework. This paper continues this line of research by integrating MCTS and minimax even more tightly into one rollout-based hybrid search algorithm, *MCTS-$\alpha\beta$*. The hybrid is able to execute two types of rollouts: MCTS rollouts and alpha-beta rollouts, i.e. rollouts implementing minimax with alpha-beta pruning and iterative deepening. During the search, all nodes accumulate both MCTS value estimates as well as alpha-beta value bounds. The two types of information are combined in a given tree node whenever alpha-beta completes a deepening iteration rooted in that node—by increasing the MCTS value estimates for the best move found by alpha-beta. A single parameter, the probability of executing MCTS rollouts vs. alpha-beta rollouts, makes it possible for the hybrid to subsume both MCTS as well as alpha-beta search as extreme cases, while allowing for a spectrum of new search algorithms in between.

Preliminary results in the game of Breakthrough show the proposed hybrid to outperform its special cases of alpha-beta and MCTS. These results are promising for the further development of rollout-based algorithms that unify MCTS and minimax approaches.

1 Introduction

Monte Carlo Tree Search (MCTS) [7,11] is a sampling-based tree search algorithm. Instead of taking all legal moves into account like traditional full-width minimax search, MCTS samples promising moves selectively. This is helpful in many large search spaces with high branching factors. Furthermore, MCTS can often take long-term effects of moves better into account than minimax, since it typically uses Monte-Carlo simulations of entire games instead of a static heuristic evaluation function for the evaluation of states. This can lead to greater positional understanding with lower implementation effort. If exploration and exploitation are traded off appropriately, MCTS asymptotically converges to the optimal policy [11], while providing approximations at any time.

© Springer International Publishing AG 2017
T. Cazenave et al. (Eds.): CGW 2016/GIGA 2016, CCIS 705, pp. 57–70, 2017.
DOI: 10.1007/978-3-319-57969-6_5

While MCTS has shown considerable success in a variety of domains [5], there are still games such as Chess and Checkers where it is inferior to minimax search with alpha-beta pruning [10]. One reason that has been identified for this weakness is the selectivity of MCTS, its focus on only the most promising lines of play. Tactical games such as Chess can have a large number of traps in their search space [16]. These can only be avoided by precise play, and the selective sampling of MCTS based on average simulation outcomes can easily miss or underestimate an important move.

In previous work [2–4], the tactical strength of minimax has been combined with the strategic and positional understanding of MCTS in *MCTS-minimax hybrids*, integrating shallow-depth minimax searches into the MCTS framework. These hybrids have shown promising results in tactical domains, both for the case where heuristic evaluation functions are unavailable [4], as well as for the case where their existence is assumed [2,3]. In this paper, we continue this line of work by integrating MCTS and minimax even more closely. Based on Huang's formulation of alpha-beta search as a rollout-based algorithm [9], we propose a hybrid search algorithm *MCTS-$\alpha\beta$* that makes use of both MCTS rollouts as well as alpha-beta rollouts. MCTS-$\alpha\beta$ can switch from executing a rollout in MCTS fashion to executing it in alpha-beta fashion at any node traversed in the tree. During the search, all nodes accumulate both MCTS value estimates as well as alpha-beta value bounds. A single rollout can collect both types of information. Whenever a deepening iteration of alpha-beta rooted in a given node is completed, the move leading to the best child found by this alpha-beta search is awarded a number of MCTS wins in that node. This allows the hybrid to combine both types of information throughout the tree.

Unlike previously proposed hybrid search algorithms, MCTS-$\alpha\beta$ subsumes both MCTS as well as alpha-beta search as extreme cases. It turns into MCTS when only using MCTS rollouts, and into alpha-beta when only using alpha-beta rollouts. By mixing both types of rollouts however, a spectrum of new search algorithms between those extremes is made available, potentially leading to better performance than either MCTS or alpha-beta in any given search domain.

This paper is structured as follows. Section 2 gives some background on MCTS and Huang's rollout-based alpha-beta as the baseline algorithms of this paper. Section 3 provides a brief overview of related work on hybrid algorithms combining features of MCTS and minimax. Section 4 outlines the proposed rollout-based hybrid MCTS-$\alpha\beta$, and Sect. 5 shows first experimental results in the test domain of Breakthrough. Section 6 finally concludes and suggests future research.

2   Background

The hybrid MCTS-$\alpha\beta$ proposed in this paper is based on two search methods as basic components: Monte Carlo Tree Search (MCTS) and minimax search with alpha-beta pruning.

2.1   MCTS

The first component of MCTS-$\alpha\beta$ is MCTS, which works by repeating the following four-phase loop until computation time runs out.

Phase one: *selection*. The tree is traversed from the root to one of its not fully expanded nodes, choosing the move to sample from each state with the help of a selection policy. The selection policy should balance the exploitation of states with high value estimates and the exploration of states with uncertain value estimates. In this paper, the popular UCT variant of MCTS is used, with the UCB1-TUNED policy as selection policy [1].

Phase two: *expansion*. When a not fully expanded node has been reached, one or more of its successors are added to the tree. In this paper, we always add the one successor chosen in the current rollout.

Phase three: *simulation*. A default policy plays the game to its end, starting from the state represented by the newly added node. MCTS converges to the optimal move in the limit even when moves are chosen randomly in this phase. Note that this phase is often also called "rollout" phase or "playout" phase in the literature. We are calling it simulation phase here, and refer to its policy as the default policy, while choosing "rollout" as the name for one entire four-phase loop. This is in order to draw a clearer connection between MCTS rollouts in this subsection and alpha-beta rollouts in the next one. It is also consistent with the terminology used in [5].

Phase four: *backpropagation*. The value estimates of all states traversed in the tree are updated with the result of the finished game.

Algorithm 1.1 shows pseudocode for a recursive formulation of MCTS used as a first starting point for this work. `gameResult(s)` returns the game-theoretic value of terminal state s. `backPropagate(s.value, score)` updates the MCTS value estimate for state s with the new result `score`. UCB1-TUNED for example requires a rollout counter, an average score and an average squared score of all previous rollouts passing through the state. Different implementations are possible for `finalMoveChoice()`; in this work, it chooses the move leading to the child of the root with the highest number of rollouts.

Many variants and extensions of this framework have been proposed in the literature [5]. In this paper, we are using MCTS with the *MCTS-Solver* extension [21] as a component of MCTS-$\alpha\beta$. MCTS-Solver is able to backpropagate not only regular simulation results such as losses and wins, but also game-theoretic values such as proven losses and proven wins whenever the search tree encounters a terminal state. The idea is marking a move as a proven loss if the opponent has a winning move from the resulting position, and marking a move as a proven win if the opponent has only losing moves from the resulting position. This avoids wasting time on the re-sampling of game states whose values are already known. Additionally, we use an informed default policy instead of a random one, making move choices based on simple knowledge about the domain at hand. It is described in Sect. 5. Both of these improvements are not essential to the idea of MCTS-$\alpha\beta$, but together allow for MCTS to win a considerable number of games against alpha-beta in our

```
1  MCTS(root) {
2      while(timeAvailable) {
3          MCTSRollout(root)
4      }
5      return finalMoveChoice()
6  }
7
8  MCTSRollout(currentState) {
9      if(currentState ∈ Tree) {
10         # selection
11         nextState ← takeSelectionPolicyMove(currentState)
12         score = MCTSRollout(nextState)
13     } else {
14         # expansion
15         addToTree(currentState)
16         # simulation
17         simulationState ← currentState
18         while(simulationState.notTerminalPosition) {
19             simulationState ← takeDefaultPolicyMove(simulationState)
20         }
21         score ← gameResult(simulationState)
22     }
23     # backpropagation
24     currentState.value ← backPropagate(currentState.value, score)
25     return score
26 }
```

Algorithm 1.1. Monte Carlo Tree Search.

experiments. This makes combining their strengths more worthwhile than if alpha-beta utterly dominated MCTS (or the other way around).

2.2 Rollout-Based Alpha-Beta

The second component of MCTS-$\alpha\beta$ is alpha-beta search. Specifically, we base this work on the rollout-based formulation of alpha-beta presented by Huang [9]. It is strictly equivalent not to classic alpha-beta search, but to an augmented version *alphabeta2*. Alphabeta2 behaves exactly like classic alpha-beta if given only one pass over the tree without any previously stored information, but it can "outprune" (evaluate fewer leaf nodes than) classic alpha-beta when called as a subroutine of a storage-enhanced search algorithm such as MT-SSS* [15]. See [9] for a detailed analysis.

The basic idea of rollout algorithms is to repeatedly start at the root and traverse down the tree. At each node representing a game state s, a selection policy chooses a successor state c from the set $C(s)$ of all legal successor states or children of s. In principle, any child could be chosen. However, it is known from alpha-beta pruning that the minimax value of the root can be determined without taking *all* children into account. Based on this realization, rollout-based alpha-beta was constructed as a rollout algorithm that restricts the selection policy at each state s to a subset of $C(s)$. This enables the algorithm to visit the same set of leaf nodes in the same order as alpha-beta, if the selection policy is chosen correctly.

Algorithm 1.2 shows pseudocode for the rollout-based formulation of alpha-beta used as a second starting point for this work. It requires the tree to maintain a closed interval $[v_s^-, v_s^+]$ for every visited state s. These intervals are initialized with $[-\infty, +\infty]$ and contain at any point in time the true minimax value of the respective state. When $v_s^- = v_s^+$, the minimax value of state s is found. When the minimax value of the root is found and the search is over, finalMoveChoice() chooses an optimal move at the root. The result of Algorithm 1.2 is independent of the implementation of takeSelectionPolicyMove(feasibleChildren); in order to achieve alpha-beta behavior however, this method always needs to return the left-most child in feasibleChildren. That is the implementation used in this work. MAX and MIN refer to states where it is the turn of the maximizing or minimizing player, respectively.

```
1  alphaBeta(root) {
2      while(v⁻_root < v⁺_root) {
3          alphaBetaRollout(root, v⁻_root, v⁺_root)
4      }
5      return finalMoveChoice()
6  }
7
8  alphaBetaRollout(s, α_s, β_s) {
9      if( C(s) ≠ ∅ ) {
10         for each c ∈ C(s) do {
11             [α_c, β_c] ← [max{α_s, v⁻_c}, min{β_s, v⁺_c}]
12         }
13         feasibleChildren ← {c ∈ C(s)|α_c < β_c}
14         nextState ← takeSelectionPolicyMove(feasibleChildren)
15         alphaBetaRollout(nextState, α_nextState, β_nextState)
16     }
```
$$v_s^- \leftarrow \begin{cases} \texttt{gameResult(s)} & \text{if } s \text{ is leaf} \\ \max_{c \in C(s)} v_c^- & \text{if } s \text{ is internal and MAX} \\ \min_{c \in C(s)} v_c^- & \text{if } s \text{ is internal and MIN} \end{cases}$$
$$v_s^+ \leftarrow \begin{cases} \texttt{gameResult(s)} & \text{if } s \text{ is leaf} \\ \max_{c \in C(s)} v_c^+ & \text{if } s \text{ is internal and MAX} \\ \min_{c \in C(s)} v_c^+ & \text{if } s \text{ is internal and MIN} \end{cases}$$
```
19 }
```

Algorithm 1.2. Rollout-based alpha-beta as proposed by Huang [9].

As mentioned by Huang, "it seems that [Algorithm 1.2] could be adapted to an 'incremental' rollout algorithm when incorporating admissible heuristic function at internal nodes (essentially an iterative deepening setting)" [9]. As shown in Sect. 4, we extended Algorithm 1.2 with an heuristic evaluation function and iterative deepening in order to create practical alpha-beta rollouts for MCTS-$\alpha\beta$. Furthermore, Huang predicted that "traditional pruning techniques and the recent Monte Carlo Tree Search algorithms, as two competing approaches for game tree evaluation, may be unified under the rollout paradigm" [9]. This is the goal of the work presented in this paper.

3   Related Work

The idea of combining the strengths of alpha-beta and MCTS in one search algorithm is motivated for instance by the work of Ramanujan et al. [16], who identified *shallow traps* as a feature of domains that are problematic for the selectively searching MCTS. Informally, Ramanujan et al. define a *level-k search trap* as the possibility of a player to choose an unfortunate move such that *after* executing the move, the opponent has a guaranteed winning strategy at most k plies deep. While such traps at shallow depths of 3 to 7 are not found in Go until the latest part of the endgame, they are relatively frequent in Chess games even at grandmaster level [16], partly explaining the success of MCTS in Go and its problems in Chess. Finnsson and Björnsson [8] discovered the similar problem of *optimistic moves*, which refers to seemingly strong moves that can be refuted right away by the opponent, but take MCTS prohibitively many simulations to evaluate correctly. The work presented in this paper is meant as a step towards search algorithms that can successfully be used in both kinds of domains—those favoring MCTS and those favoring alpha-beta until now.

Previous work on developing algorithms influenced by both MCTS and minimax has taken two main approaches. The first approach is to embed minimax searches within the MCTS framework. Shallow minimax searches have for example been used in every step of the simulation phase for Lines of Action [20], Chess [17], and various multi-player games [14]. Baier and Winands studied approaches that use minimax search without evaluation functions nested into the selection/-expansion phase, the simulation phase, and the backpropagation phase of MCTS [4], as well as approaches that use minimax search with evaluation functions in the simulation phase, for early termination of simulations, and as a prior for tree nodes [2,3].

The second approach is to identify individual features of minimax such as minimax-style backups, and integrate them into MCTS. In the algorithm $UCTMAX_H$ [18] for example, MCTS simulations are replaced with heuristic evaluations and classic averaging MCTS backups with minimaxing backups. In *implicit minimax backups* [12], both minimaxing backups of heuristic evaluations and averaging backups of simulation returns are managed simultaneously.

This paper takes a new approach. While in our previous hybrids [2–4], alpha-beta searches were nested into the MCTS framework and had to complete before MCTS could continue—MCTS and alpha-beta functioned as combined, but separate algorithms—the newly proposed MCTS-$\alpha\beta$ tightly interleaves MCTS and alpha-beta. The formulation of alpha-beta as a rollout algorithm [9] allows MCTS-$\alpha\beta$ to decide about continuing a rollout in MCTS fashion or in alpha-beta fashion at every node encountered during the search. As opposed to $UCTMAX_H$ and *implicit minimax* mentioned above, MCTS-$\alpha\beta$ is not picking and combining individual features of MCTS and minimax. It subsumes both regular MCTS and regular alpha-beta when only MCTS rollouts or only alpha-beta rollouts are used, but results in a new type of search algorithm when both types are combined. A probability parameter p determines the mix.

Apart from Huang [9], several other researchers have proposed rollout-based formulations of minimax search. For example, Weinstein, Littman, and Goschin [19] presented a rollout algorithm that outprunes alpha-beta, and Chen et al. [6] proposed a rollout algorithm similar to MT-SSS*. We chose Huang's alpha-beta formulation as a basis for this work because of its clear formal characterization, unifying both alpha-beta and MT-SSS* under the rollout framework.

4   MCTS-$\alpha\beta$

The basic idea of MCTS-$\alpha\beta$ is to allow for a mix of MCTS and alpha-beta rollouts. A simple way of achieving this is by introducing a parameter $p \in [0, 1]$ as the probability of starting an MCTS rollout at the root. $1 - p$, conversely, is the probability of starting an alpha-beta rollout instead. Assume that an MCTS rollout is chosen at the root. At every recursive call of the MCTS rollout, the randomized decision is made again whether to continue with MCTS or whether to switch to alpha-beta, using the same probabilities. If the search tree is left without switching to an alpha-beta rollout at any point, the simulation and backpropagation phases are executed just like in a regular MCTS rollout. MCTS value estimates are updated in all traversed nodes, and the next rollout begins.

If however any randomized decision indicates the start of an alpha-beta rollout—either at the root or at a later stage of an MCTS rollout—then the rollout continues in alpha-beta fashion, with the current node functioning as the root of the alpha-beta search. This is similar to starting an embedded alpha-beta search at the current node, like the MCTS-IP-M algorithm described in [2,3]. But MCTS-$\alpha\beta$ does not necessarily execute the entire alpha-beta search. The newly proposed hybrid can execute only one alpha-beta rollout instead, and potentially continue this particular alpha-beta search at any later point during the search process—whenever the decision for an alpha-beta rollout is made again at the same node. The interleaving of MCTS and minimax is more fine-grained than in previous hybrids.

There are a few differences between the alpha-beta rollouts of Huang's Algorithm 1.2 and those of MCTS-$\alpha\beta$. First, MCTS-$\alpha\beta$ uses an evaluation function to allow for depth-limited search. Second, these depth-limited searches are conducted in an iterative deepening manner. Third, MCTS-$\alpha\beta$ can reduce the branching factor for alpha-beta rollouts with the help of move ordering and k-best pruning (only searching the k moves ranked highest by a move ordering function). Fourth, if an alpha-beta rollout of MCTS-$\alpha\beta$ was called from an ongoing MCTS rollout instead of from the root, it returns the evaluation value of its leaf node to that MCTS rollout for backpropagation. And fifth, if the alpha-beta rollout is finishing a deepening iteration in a state s—if it is completing a 1-ply, 2-ply, 3-ply search etc—MCTS-$\alpha\beta$ gives a bonus in MCTS rollouts to the MCTS value estimate of the best child of s found in that iteration. At the same time, the bonus given for the previous deepening iteration is removed, so that only the currently best child of s is boosted.

This last point makes it clear how MCTS-$\alpha\beta$ can subsume both MCTS and alpha-beta. If $p = 1$, only MCTS rollouts are executed, and MCTS-$\alpha\beta$ behaves exactly like regular MCTS. If $p = 0$, only alpha-beta rollouts are started immediately at the root, and only the best move found by the last deepening iteration has a positive MCTS value estimate due to its bonus. MCTS-$\alpha\beta$ therefore behaves exactly like alpha-beta (an iterative deepening version of Huang's augmented alphabeta2, to be precise). If $0 < p < 1$ however, MCTS-$\alpha\beta$ becomes a true hybrid, combining MCTS and minimax behavior throughout the search tree, and choosing moves at the root based on both real MCTS rollout counts as well as MCTS rollout bonuses from the last completed deepening iteration of alpha-beta.

Algorithm 1.3 shows pseudocode for MCTS-$\alpha\beta$. $D(s)$ is the current search depth for alpha-beta starting in state s, initialized to 1 for all states. $K(s)$ is the set of the k best successor states or children of state s as determined by the move ordering function. `random(0,1)` returns a random, uniformly distributed value in $[0,1]$. $\left[v_{s,d}^-, v_{s,d}^+\right]$ is an interval containing the value of state s when searched to depth d. `eval(s)` is a heuristic evaluation of state s, and `sigmoid(x)` is a sigmoid transformation used to spread out heuristic evaluations to the interval $[0,1]$. `s.giveBonus(b)` adds b winning rollouts to the MCTS value estimate of state s, and `s.removeLastBonusGiven()` removes the last bonus given to s. `finalMoveChoice()` is the same as for regular MCTS, choosing the child with the most rollouts at the root.

The parameters of MCTS-$\alpha\beta$ are the MCTS rollout probability p, the bonus weight w, the bonus weight factor f that defines how much stronger bonuses become with the depth of the completed alpha-beta search, the number of moves k for k-best pruning, and the maximum minimax depth l. When a depth-l alpha-beta search starting from state s is completed, the search depth is not further increased there. Only MCTS rollouts will be started from s in the rest of the search time.

Note that while an MCTS rollout can turn into an alpha-beta rollout at any node, a mid-rollout switch from alpha-beta back to MCTS is not possible in MCTS-$\alpha\beta$.

5   Experimental Results

We conducted preliminary experiments with MCTS-$\alpha\beta$ in the deterministic, two-player, zero-sum game of *Breakthrough*. MCTS parameters such as the exploration factor C ($C = 0.8$) were optimized for the baseline MCTS-Solver and kept constant during testing. We used minimax with alpha-beta pruning, move ordering, k-best pruning, and iterative deepening, but no other search enhancements. Every experimental condition consisted of 1000 games, with each player playing 500 as White and 500 as Black. All algorithms were allowed to expand 2500 nodes before making each move decision, unless specified otherwise. A node

```
 1  MCTSAlphaBeta(root) {
 2      while(timeAvailable) {
 3          if(random(0,1)<p) {
 4              MCTSRollout(root)
 5          } else {
 6              alphaBetaRollout(root, D(root), v⁻_root,D(root), v⁺_root,D(root))
 7          }
 8      }
 9      return finalMoveChoice()
10  }
11
12  alphaBetaRollout(s, d, α_s, β_s) {
13      if( K(s) ≠ ∅ and d > 0 ) {
14          for each c ∈ K(s) do {
15              [α_c, β_c] ← [max{α_s, v⁻_c,d-1}, min{β_s, v⁺_c,d-1}]
16          }
17          feasibleChildren ← {c ∈ K(s)|α_c < β_c}
18          nextState ← takeSelectionPolicyMove(feasibleChildren)
19          rolloutResult ← alphaBetaRollout(nextState, d-1, α_nextState, β_nextState)
20      }
```

$$
v^-_{s,d} \leftarrow \begin{cases}
\text{gameResult}(s) & \text{if } s \text{ is leaf} \\
\text{eval}(s) & \text{if } d = 0 \\
\max_{c \in K(s)} v^-_{c,d-1} & \text{if } d > 0 \text{ and } s \text{ is internal and MAX} \\
\min_{c \in K(s)} v^-_{c,d-1} & \text{if } d > 0 \text{ and } s \text{ is internal and MIN}
\end{cases}
$$

$$
v^+_{s,d} \leftarrow \begin{cases}
\text{gameResult}(s) & \text{if } s \text{ is leaf} \\
\text{eval}(s) & \text{if } d = 0 \\
\max_{c \in K(s)} v^+_{c,d-1} & \text{if } d > 0 \text{ and } s \text{ is internal and MAX} \\
\min_{c \in K(s)} v^+_{c,d-1} & \text{if } d > 0 \text{ and } s \text{ is internal and MIN}
\end{cases}
$$

```
23      if(K(s) = ∅ or d = 0) {
24          rolloutResult ← v⁺_s,d
25      }
26      return rolloutResult
27  }
28
29  MCTSRollout(currentState) {
30      if(currentState ∈ Tree) {
31          nextState ← takeSelectionPolicyMove(currentState)
32          if(random(0,1)<p or D(nextState)= l {
33              score ← MCTSRollout(nextState)
34          } else {
35              score ← alphaBetaRollout(nextState, D(nextState),
36                  v⁻_nextState,D(nextState), v⁺_nextState,D(nextState))
37              score ← sigmoid(score)
38              if(v⁻_nextState,D(nextState) = v⁺_nextState,D(nextState)) {
39                  bestChild ← bestChildFoundByAlphaBetaIn(nextState)
40                  bonus ← score * w * f^D(nextState)
41                  bestChild.removeLastBonusGiven()
42                  bestChild.giveBonus(bonus)
43                  D(nextState) ← D(nextState)+1
44              }
45          }
46      } else {
47          addToTree(currentState)
48          simulationState ← currentState
49          while(simulationState.notTerminalPosition) {
50              simulationState ← takeDefaultPolicyMove(simulationState)
51          }
52          score ← gameResult(simulationState)
53      }
54      currentState.value ← backPropagate(currentState.value, score)
55      return score
56  }
```

Algorithm 1.3. MCTS-$\alpha\beta$.

limit was chosen instead of a time limit in order to first test whether the newly proposed hybrid can search more effectively than its special cases MCTS and alpha-beta, without taking into account the additional questions of using more or less computationally expensive evaluation functions and MCTS default policies.

Subsection 5.1 describes the rules of Breakthrough as well as the evaluation function, move ordering, and default policy used for it. Subsection 5.2 shows the results.

5.1   Test Domain

The variant of Breakthrough used in our experiments is played on a 6×6 board. The game was originally described as being played on a 7×7 board, but other sizes such as 8×8 are popular as well, and the 6×6 board preserves an interesting search space.

At the beginning of the game, White occupies the first two rows of the board, and Black occupies the last two rows of the board. The two players alternatingly move one of their pieces straight or diagonally forward. Two pieces cannot occupy the same square. However, players can capture the opponent's pieces by moving diagonally onto their square. The game is won by the player who succeeds first at advancing one piece to the home row of her opponent, i.e. reaching the first row as Black or reaching the last row as White.

The simple evaluation function we use for Breakthrough gives the player one point for each piece of her color. The opponent's points are subtracted, and the resulting value is then normalized to the interval $[0, 1]$.

The move ordering ranks winning moves first. Second, it ranks saving moves (captures of an opponent piece that is only one move away from winning). Third, it ranks captures, and fourth, all other moves. Within all four groups of moves, moves that are closer to the opponent's home row are preferred. When two moves are ranked equally by the move ordering, they are searched in random order.

The informed default policy used for the experiments always chooses the move ranked first by the above move ordering function.

5.2   Performance of MCTS-$\alpha\beta$

In our first set of experiments, we hand-tuned the five MCTS-$\alpha\beta$ parameters against two opponents: regular alpha-beta with the same evaluation function, move ordering, and k-best pruning ($k = 10$ was found to be the strongest setting, which confirms our observations in [2] for 6×6 Breakthrough), as well as regular MCTS with the same informed default policy as described in the previous subsection. The best parameter settings found were $k = 8$, $l = 6$, $w = 200$, $f = 8$, and $p = 0.95$. With these settings, the results for MCTS-$\alpha\beta$ were a winrate of 63.7% against alpha-beta and 58.2% against MCTS. This means MCTS-$\alpha\beta$ is significantly stronger than both of its basic components ($p < 0.001$). MCTS won 63.6% of 1000 games against alpha-beta, which means that also in the resulting round-robin competition between the three players MCTS-$\alpha\beta$ performed best with 1219 won games, followed by MCTS with 1054 and alpha-beta with 727 wins in total. The optimal parameters for MCTS-$\alpha\beta$ in this scenario include a high MCTS rollout probability ($p = 0.95$), resulting in few alpha-beta rollouts being carried out. This fact seems to be agree with the strong performance of regular MCTS against regular alpha-beta. MCTS rollouts with an informed default policy seem to be more effective than alpha-beta rollouts with the primitive evaluation function described above, especially when the higher computational cost of the MCTS simulations is not taken into consideration.

In a second set of experiments, we contrasted this with a different scenario where no strong default policy is available. Instead, MCTS simulations are stopped after 3 random moves, the heuristic evaluation function is applied to the current state, and the resulting value is backpropagated as MCTS simulation return. This technique is called MCTS-IC here for consistency with previous work [3]; a similar technique where the evaluation function value is rounded to either a win or a loss before backpropagation has also been studied under the name MCTS-EPT [13]. Alpha-beta rollouts remain unchanged. MCTS-IC with the simple evaluation function we are using is weaker than MCTS with the strong informed policy described above—in a direct comparison, MCTS-IC won only 20.5% of 1000 games. In this setting, and keeping all other parameters constant, MCTS-$\alpha\beta$ performed best with $p = 0.3$. This confirms that as soon as MCTS is weakened in comparison to alpha-beta, alpha-beta rollouts become more effective for MCTS-$\alpha\beta$ than MCTS rollouts, and the optimal value for p is lower. The results of 1000 games against the baselines were 53.7% against regular alpha-beta (not significantly different), and 73.1% against MCTS-IC (here the hybrid is significantly stronger with $p < 0.001$). With alpha-beta winning 73.1% of games against MCTS-IC as well, this means a round-robin result of 1268 wins for MCTS-$\alpha\beta$, now followed by alpha-beta with 1194 and MCTS-IC with 538 wins. Although the lack of a strong MCTS default policy has pushed alpha-beta ahead of MCTS, the hybrid algorithm still leads.

Figures 1 and 2 illustrate the performance landscape of MCTS-$\alpha\beta$ with regard to the crucial p parameter, both in the scenario with informed simulations and in the scenario with MCTS-IC simulations. Each data point results from 1000 games against regular alpha-beta.

In a third and last set of experiments, we tested the generalization of the behavior of MCTS-$\alpha\beta$ to different time settings. In the scenario with MCTS-IC simulations, all parameter settings were left unchanged ($p = 0.3$, $k = 8$, $l = 6$, $w = 200$, $f = 8$), but all algorithms were now allowed 10000 nodes per move. MCTS-$\alpha\beta$ won 51.9% of 1000 games against regular alpha-beta, and 76.4% against regular MCTS-IC. Alpha-beta won 71.9% of games against MCTS-IC. The round-robin result is 1283 wins for MCTS-$\alpha\beta$, followed by alpha-beta with 1200 and MCTS-IC with 517 wins. The algorithms were also tested against each other with only 500 nodes per move—here parameter l was reduced to 4, while all other parameters stayed the same (experience with other MCTS-minimax hybrids has shown that shorter search times often profit from keeping alpha-beta more shallow [2]). For this setting, the results were 54.6% for MCTS-$\alpha\beta$

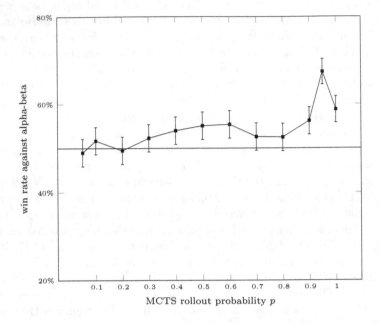

Fig. 1. Performance of MCTS-$\alpha\beta$ with informed MCTS simulations.

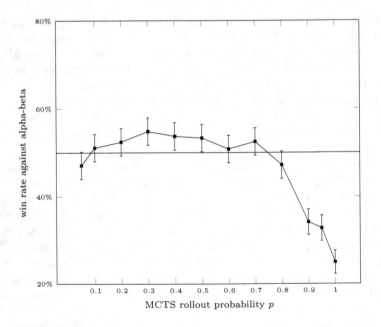

Fig. 2. Performance of MCTS-$\alpha\beta$ with MCTS-IC simulations.

versus alpha-beta, 67.2% for MCTS-$\alpha\beta$ versus MCTS-IC, and 74.0% for alpha-beta versus MCTS-IC. Added up, this results in a round-robin with 1218 wins for MCTS-$\alpha\beta$, 1194 for alpha-beta, and 588 for MCTS-IC. In conclusion, the relative performance of MCTS-$\alpha\beta$ generalized to time settings 4 times longer as well as 5 times shorter without requiring extensive re-tuning.

6   Conclusion and Future Research

In this paper, we introduced the new hybrid search algorithm MCTS-$\alpha\beta$. It is based on MCTS rollouts and alpha-beta rollouts and unifies both search approaches under the same framework. While subsuming regular alpha-beta and regular MCTS as extreme cases, MCTS-$\alpha\beta$ opens a new space of search algorithms in between.

Preliminary results in the game of Breakthrough are promising, but do not constitute much more than a proof of concept yet. More work has to be done to gain an understanding of MCTS-$\alpha\beta$, and to further develop rollout-based MCTS-minimax hybrids. A first possible research direction is the exploration of different design choices in the algorithm. Can alpha-beta and MCTS rollouts be more intelligently combined than by choosing them at random? How much playing strength comes from the backpropagated evaluation values, and how much from the MCTS bonuses given after alpha-beta finishes a search depth? A second direction is an analysis of the conditions under which MCTS-$\alpha\beta$ works best. Does it only show promise when the performance of MCTS and alpha-beta in the domain at hand are at least roughly comparable, or can it also improve an algorithm which is already clearly superior? How does MCTS-$\alpha\beta$ perform against MCTS and alpha-beta at equal time controls? And finally, a comparison of MCTS-$\alpha\beta$ with previously proposed hybrids would be of great interest.

Acknowledgment. The author thanks the Games and AI group, Department of Data Science and Knowledge Engineering, Maastricht University, for computational support.

References

1. Auer, P., Cesa-Bianchi, N., Fischer, P.: Finite-time analysis of the multiarmed bandit problem. Mach. Learn. **47**(2–3), 235–256 (2002)
2. Baier, H.: Monte-Carlo Tree Search Enhancements for One-Player and Two-Player Domains. Ph.D. thesis, Maastricht University, Maastricht, The Netherlands (2015)
3. Baier, H., Winands, M.H.M.: Monte-Carlo tree search and minimax hybrids with heuristic evaluation functions. In: Cazenave, T., Winands, M.H.M., Björnsson, Y. (eds.) CGW 2014. CCIS, vol. 504, pp. 45–63. Springer, Cham (2014). doi:10.1007/978-3-319-14923-3_4
4. Baier, H., Winands, M.H.M.: MCTS-minimax hybrids. IEEE Trans. Comput. Intell. AI Games **7**(2), 167–179 (2015)
5. Browne, C., Powley, E.J., Whitehouse, D., Lucas, S.M., Cowling, P.I., Rohlfshagen, P., Tavener, S., Perez, D., Samothrakis, S., Colton, S.: A survey of Monte Carlo tree search methods. IEEE Trans. Comput. Intell. AI Games **4**(1), 1–43 (2012)

6. Chen, J., Wu, I., Tseng, W., Lin, B., Chang, C.: Job-level alpha-beta search. IEEE Trans. Comput. Intell. AI Games **7**(1), 28–38 (2015)

7. Coulom, R.: Efficient selectivity and backup operators in Monte-Carlo tree search. In: van den Herik, H.J., Ciancarini, P., Donkers, H.H.L.M.J. (eds.) CG 2006. LNCS, vol. 4630, pp. 72–83. Springer, Heidelberg (2007). doi:10.1007/978-3-540-75538-8_7

8. Finnsson, H., Björnsson, Y.: Game-tree properties and MCTS performance. In: IJCAI 2011 Workshop on General Intelligence in Game Playing Agents (GIGA 2011), pp. 23–30 (2011)

9. Huang, B.: Pruning game tree by rollouts. In: Bonet, B., Koenig, S. (eds.) Twenty-Ninth AAAI Conference on Artificial Intelligence, AAAI 2015, pp. 1165–1173. AAAI Press (2015)

10. Knuth, D.E., Moore, R.W.: An analysis of alpha-beta pruning. Artif. Intell. **6**(4), 293–326 (1975)

11. Kocsis, L., Szepesvári, C.: Bandit based Monte-Carlo planning. In: Fürnkranz, J., Scheffer, T., Spiliopoulou, M. (eds.) ECML 2006. LNCS (LNAI), vol. 4212, pp. 282–293. Springer, Heidelberg (2006). doi:10.1007/11871842_29

12. Lanctot, M., Winands, M.H.M., Pepels, T., Sturtevant, N.R.: Monte Carlo tree search with heuristic evaluations using implicit minimax backups. In: 2014 IEEE Conference on Computational Intelligence and Games, CIG 2014, pp. 341–348. IEEE (2014)

13. Lorentz, R.: Early Playout Termination in MCTS. In: Plaat, A., van den Herik, J., Kosters, W. (eds.) ACG 2015. LNCS, vol. 9525, pp. 12–19. Springer, Cham (2015). doi:10.1007/978-3-319-27992-3_2

14. Nijssen, J.P.A.M., Winands, M.H.M.: Playout search for Monte-Carlo tree search in multi-player games. In: van den Herik, H.J., Plaat, A. (eds.) ACG 2011. LNCS, vol. 7168, pp. 72–83. Springer, Heidelberg (2012). doi:10.1007/978-3-642-31866-5_7

15. Plaat, A., Schaeffer, J., Pijls, W., de Bruin, A.: Best-first fixed-depth minimax algorithms. Artif. Intell. **87**(1), 255–293 (1996)

16. Ramanujan, R., Sabharwal, A., Selman, B.: On adversarial search spaces and sampling-based planning. In: Brafman, R.I., Geffner, H., Hoffmann, J., Kautz, H.A. (eds.) 20th International Conference on Automated Planning and Scheduling, ICAPS 2010, pp. 242–245. AAAI (2010)

17. Ramanujan, R., Sabharwal, A., Selman, B.: Understanding sampling style adversarial search methods. In: Grünwald, P., Spirtes, P. (eds.) 26th Conference on Uncertainty in Artificial Intelligence, UAI 2010, pp. 474–483 (2010)

18. Ramanujan, R., Selman, B.: Trade-offs in sampling-based adversarial planning. In: Bacchus, F., Domshlak, C., Edelkamp, S., Helmert, M. (eds.) 21st International Conference on Automated Planning and Scheduling, ICAPS 2011. AAAI (2011)

19. Weinstein, A., Littman, M.L., Goschin, S.: Rollout-based game-tree search outprunes traditional alpha-beta. In: Deisenroth, M.P., Szepesvári, C., Peters, J. (eds.) JMLR Proceedings Tenth European Workshop on Reinforcement Learning, EWRL 2012, vol. 24, pp. 155–167 (2012)

20. Winands, M.H.M., Björnsson, Y.: Alpha-beta-based play-outs in Monte-Carlo tree search. In: Cho, S.B., Lucas, S.M., Hingston, P. (eds.) 2011 IEEE Conference on Computational Intelligence and Games, CIG 2011, pp. 110–117. IEEE (2011)

21. Winands, M.H.M., Björnsson, Y., Saito, J.-T.: Monte-Carlo tree search solver. In: van den Herik, H.J., Xu, X., Ma, Z., Winands, M.H.M. (eds.) CG 2008. LNCS, vol. 5131, pp. 25–36. Springer, Heidelberg (2008). doi:10.1007/978-3-540-87608-3_3

Learning from the Memory of Atari 2600

Jakub Sygnowski[(✉)] and Henryk Michalewski

Department of Mathematics, Informatics, and Mechanics,
University of Warsaw, Warsaw, Poland
J.Sygnowski@students.mimuw.edu.pl, H.Michalewski@mimuw.edu.pl

Abstract. We train a number of neural networks to play the games
Bowling, Breakout and Seaquest using information stored in the memory of a video game console Atari 2600. We consider four models of neural
networks which differ in size and architecture: two networks which use
only information contained in the RAM and two mixed networks which
use both information in the RAM and information from the screen.

As the benchmark we used the convolutional model proposed in [17]
and received comparable results in all considered games. Quite surprisingly, in the case of Seaquest we were able to train RAM-only agents
which behave better than the benchmark screen-only agent. Mixing
screen and RAM did not lead to an improved performance comparing to
screen-only and RAM-only agents.

1 Introduction

An Atari 2600 controller can perform one of 18 actions[1] and in this work we are
intended to learn which of these 18 actions is the most profitable given a state of
the screen or memory. Our work is based on deep Q-learning [17] – a reinforcement learning algorithm that can learn to play Atari games using only input
from the screen. The deep Q-learning algorithm builds on the Q-learning [25]
algorithm, which in its simplest form (see [19, Fig. 21.8]) iteratively learns values
$Q(state, action)$ for *all* state-action pairs and lets the agent choose the action
with the highest value. In the instance of Atari 2600 games this implies evaluating
all pairs $(screen, action)$ where $action$ is one of the 18 positions of the controller.
This task is infeasible and similarly, learning *all* values $Q(state, action)$ is not a
realistic task in other real-world games such as chess or Go.

This feasibility issues led to generalizations of the Q-learning algorithm which
are intended to limit the number of parameters on which the function Q depends.
One can arbitrarily restrict the number of features which can be learned[2], but

---

[1] For some games only some of these 18 actions are used in the gameplay. The number
of available actions is 4 for Breakout, 18 for Seaquest and 6 for Bowling.

[2] E.g. we may declare that $Q(state, action) = \theta_1 f_1(state, action) + \theta_2 f_2(state, action)$,
where f_1, f_2 are some fixed pre-defined functions, for example f_1 may declare value
1 to the state-action pair $(screen, fire)$ if a certain shape appears in the bottom-left
corner of the screen and 0 otherwise and f_2 may declare value 1 to $(screen, left)$
if an enemy appeared on the right and 0 otherwise. Then the Q-learning algorithm
learns the best values of θ_1, θ_2.

© Springer International Publishing AG 2017
T. Cazenave et al. (Eds.): CGW 2016/GIGA 2016, CCIS 705, pp. 71–85, 2017.
DOI: 10.1007/978-3-319-57969-6_6

instead of using manually devised features, the deep Q-learning algorithm[3] presented in [17] builds them in the process of training of the neural network. Since every neural network is a composition of a priori unknown linear maps and fixed non-linear maps, the aim of the deep Q-learning algorithm is to learn coefficients of the unknown linear maps.

In the deep Q-learning algorithm the game states, actions and immediate rewards are passed to a *deep convolutional network*. This type of network abstracts features of the screen, so that various patterns on the screen can be identified as similar. The network has a number of output nodes – one for each possible action – and it predicts the cumulative game rewards after making moves corresponding to actions.

A number of decisions was made in the process of designing of the deep Q-learning algorithm (see [17, Algorithm 1] for more details): (1) in each step there is some probability ε of making a random move and it decreases from $\varepsilon = 1$ to $\varepsilon = 0.1$ in the course of training, (2) the previous game states are stored in the *replay memory*; the updates in the Q-learning are limited to a random batch of events polled from that memory, (3) the updates of unknown linear maps in the neural network are performed according to the gradient of the squared loss function which measures discrepancy between the estimation given by the network and the actual reward. In this work we use the same computational infrastructure as in [17], including the above decisions (1)–(3).

Related Work

The original algorithm in [17] was improved in a number of ways in [15,18,21]. This includes changing network architecture, choosing better hyperparameters and improving the speed of algorithm which optimizes neural network's loss function. These attempts proved to be successful and made the deep Q-learning algorithm the state-of-the-art method for playing Atari games.

Instead of the screen one can treat the RAM state of the Atari machine as the game state. The work [16] implemented a classical planning algorithm on the RAM state. Since the Atari 2600 RAM consists of only 128 bytes, one can efficiently search in this low-dimensional state space. Nevertheless, the learning in [16] happens during the gameplay, so it depends on the time limit for a single move. In contrast, in [17] the learning process happens before the gameplay - in particular the agent can play in the real-time. To the best of our knowledge the only RAM-based agent not depending on search was presented in [11]. We cite these results as ale_ram.

In our work we use the deep Q-learning algorithm, but instead of using screens as inputs to the network, we pass the RAM state or the RAM state and the screen together. In the following sections we describe the games we used for evaluation, as well as the network architectures we tried and hyperparameters we tweaked.

---

[3] This algorithm is also called a deep q-network or DQN.

The Work [17] as the Main Benchmark

The changes to the deep Q-learning algorithm proposed in [18] came at a cost of making computations more involved comparing to [17]. In this work we decided to use as the reference result only the basic work [17], which is not the state of the art, but a single training of a neural network can be contained in roughly 48 hours using the experimental setup we describe below. This decision was also motivated by a preliminary character of our study – we wanted to make sure that indeed the console memory contains useful data which can be extracted during the training process using the deep Q-learning algorithm. From this perspective the basic results in [17] seem to be a perfect benchmark to verify feasibility of learning from RAM. We refer to this benchmark architecture as nips through this paper.

2   The Setting of the Experiment

2.1   Games

Bowling – simulation of the game of bowling; the player aims the ball toward the pins and then steers the ball; the aim is to hit the pins [1,2].
Breakout – the player bounces the ball with the paddle towards the layer of bricks; the task is to destroy all bricks; a brick is destroyed when the ball hits it [3,4].
Seaquest – the player commands a submarine, which can shoot enemies and rescue divers by bringing them above the water-level; the player dies if he fails to get a diver up before the air level of submarine vanishes [8,9].

We've chosen these games, because each of them offers a distinct challenge. Breakout is a relatively easy game with player's actions limited to moves along the horizontal axis. We picked Breakout because disastrous results of learning would indicate a fundamental problem with the RAM learning. The deep Q-network for Seaquest constructed in [17] plays at an amateur human level and for this reason we consider this game as a tempting target for improvements. Also the game state has some elements that possibly can be detected by the RAM-only network (e.g. oxygen-level meter or the number of picked divers). Bowling

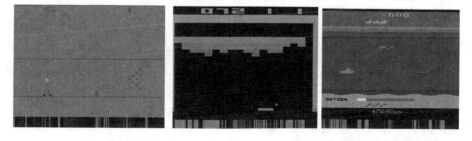

Fig. 1. From left to right Bowling, Breakout and Seaquest. The 128 vertical bars and the bottom of every screenshot represent the state of the memory, with black representing 0 and lighter color corresponding to higher values of a given memory cell.

seems to be a hard game for all deep Q-network models. It is an interesting target for the RAM-based networks, because visualizations suggest that the state of the RAM is changing only very slightly (Fig. 1).

2.2   Technical Architecture

By one experiment we mean a complete training of a single deep Q-network. In this paper we quote numerical outcomes of 30 experiments which we performed[4]. For our experiments we made use of Nathan Sprague's implementation of the deep Q-learning algorithm [6] in Theano [12] and Lasagne [5]. The code uses the Arcade Learning Environment [11] – the standard framework for evaluating agents playing Atari games. Our code with instructions how to run it can be found on github [7]. All experiments were performed on a Linux machine equipped with Nvidia GTX 480 graphics card. Each of the experiments lasted for 1–3 days. A single epoch of a RAM-only training lasted approximately half of the time of the screen-only training for an architecture with the same number of layers.

2.3   Network Architectures

We performed experiments with four neural network architectures which accept the RAM state as (a part of) the input. The RAM input is scaled by 256, so all the inputs are between 0 and 1.

All the hyperparameters of the network we consider are the same as in [17], if not mentioned otherwise (see Appendix A). We only changed the size of the replay memory to $\approx 10^5$ items, so that it fits into 1.5 GB of Nvidia GTX 480 memory[5].

3   Plain Approach

Here we present the results of training the RAM-only networks just_ram and big_ram as well as the benchmark model nips.

Neural network 1. just_ram(outputDim)

Input: RAM
Output: A vector of length outputDim
1   $hiddenLayer1 \leftarrow DenseLayer(RAM, 128, rectify)$
2   $hiddenLayer2 \leftarrow DenseLayer(hiddenLayer1, 128, rectify)$
3   $output \leftarrow DenseLayer(hiddenLayer2, outputDim, no\ activation)$
4   **return** $output$

The next considered architecture consists of the above network with two additional dense layers:

---

[4] The total number of experiments exceeded 100, but this includes experiments involving other models and repetitions of experiments described in this paper.
[5] We have not observed a statistically significant change in results when switching between replay memory size of 10^5 and $5 \cdot 10^5$.

Neural network 2. big_ram(outputDim)

Input: RAM
Output: A vector of length outputDim

1 $hiddenLayer1 \leftarrow DenseLayer(RAM, 128, rectify)$
2 $hiddenLayer2 \leftarrow DenseLayer(hiddenLayer1, 128, rectify)$
3 $hiddenLayer3 \leftarrow DenseLayer(hiddenLayer2, 128, rectify)$
4 $hiddenLayer4 \leftarrow DenseLayer(hiddenLayer3, 128, rectify)$
5 $output \leftarrow DenseLayer(hiddenLayer4, outputDim, no\ activation)$
6 **return** $output$

The training process consists of *epochs*, which are interleaved by test periods. During a test period we run the model with the current parameters, the probability of doing a random action $\varepsilon = 0.05$ and the number of test steps (frames) being 10 000. Figures 2, 3 and 4 show the average result per *episode* (full game, until player's death) for each epoch.

Figures 2, 3, and 4 show that there is a big variance of the results between epochs, especially in the RAM models. Because of that, to compare the models, we chose the results of the best epoch[6]. We summarized these results in Table 1, which also include the results of ale_ram[7].

In Breakout the best result for the big_ram model is weaker than those obtained by the network nips. In Seaquest the best result obtained by the big_ram network is better than the best result obtained by the network nips.

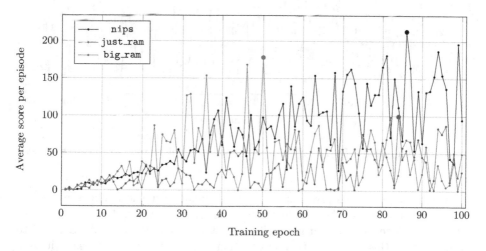

Fig. 2. Training results for Breakout for three plain models: nips, just_ram, big_ram.

---

[6] For Breakout we tested networks with best training-time results. The test consisted of choosing other random seeds and performing 100 000 steps. For all networks, including nips, we received results consistently lower by about 30%.

[7] The ale_ram's evaluation method differ – the scores presented are the average over 30 trials consisting of a long period of learning and then a long period of testing, nevertheless the results are much worse than of any DQN-based method presented here.

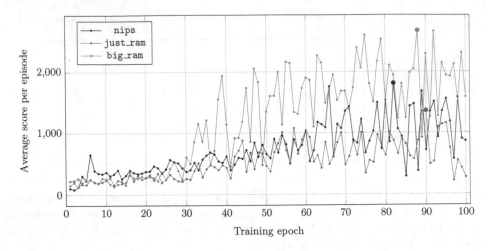

Fig. 3. Training results for Seaquest for three plain models: `nips`, `just_ram`, `big_ram`.

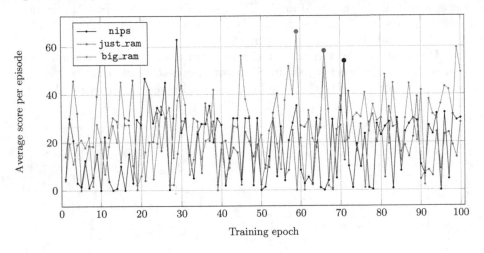

Fig. 4. Training results for Bowling for three plain models: `nips`, `just_ram`, `big_ram`.

In Bowling our methods give a slight improvement over the network `nips`, yet in all considered approaches the learning as illustrated by Fig. 4 seem to be poor and the outcome in terms of gameplay is not very satisfactory. We decided to not include in this paper further experiments with Bowling and leave it as a topic of a further research.

4   Regularization

Training a basic RAM-only network leads to high variance of the results (see the figures in the previous section) over epochs. This can be a sign of overfitting.

Table 1. Table summarizing test results for basic methods.

	Breakout	Seaquest	Bowling
nips best	**213.14**	1808	54.0
just_ram best	99.56	1360	58.25
big_ram best	178.0	**2680**	**66.25**
ale_ram	4.0	593.7	29.3

To tackle this problem we have applied *dropout* [20], a standard regularization technique for neural networks.

Dropout is a simple, yet effective regularization method. It consists of "turning off" with probability p each neuron in training, i.e. setting the output of the neuron to 0, regardless of its input. In backpropagation, the parameters of switched off nodes are not updated. Then, during testing, all neurons are set to "on" – they work as in the course of normal training, with the exception that each neuron's output is multiplied by p to make up for the skewed training. The intuition behind the dropout method is that it forces each node to learn in absence of other nodes. The work [24] shows experimental evidence that the dropout method indeed reduces the variance of the learning process.

We've enabled dropout with probability of turning off a neuron $p = \frac{1}{2}$. This applies to all nodes, except output ones. We implemented dropout for two RAM-only networks: just_ram and big_ram. This method offers an improvement for the big_ram network leading to the best result for Seaquest in this paper. The

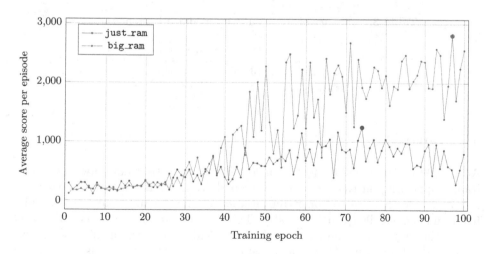

Fig. 5. Training results for Seaquest with dropout $p = 0.5$ for models just_ram, big_ram. This figure suggests that indeed dropout reduces the variance of the learning process.

Table 2. Summary of test results for training which involves regularization with the dropout method with the parameter $p = 0.5$.

	Breakout	Seaquest
just_ram with dropout best	130.5	1246.67
big_ram with dropout best	122.25	**2805**

best epoch results are presented in the Table 2 and the intermediate training results for Seaquest are shown in Fig. 5.

5   Decreasing Learning Rate

We also tried to reduce the variance of the learner through reduction of the learning rate from 0.0002 to 0.0001.

The learning rate is a parameter of the algorithm *rmsprop* that decides how parameters are changed in each step. Bigger learning rates correspond to moving faster in the parameter space, making learning faster, but more noisy.

We expected that the drastic changes in performance between consecutive epochs, as illustrated by Figs. 2 and 3, may come from stepping over optimal values when taking too big steps. If it is the case, decreasing the step size should lead to slower learning combined with higher precision of finding minima of the loss function.

The results of these experiments can be found in Table 3. Comparing to the training without regularization, scores improved only in the case of Breakout and the just_ram network, but not by a big margin.

Table 3. Summary of test results for modified learning rate.

	Breakout	Seaquest
just_ram best	137.67	1233.33
big_ram best	112.14	2675

6   Frame Skip

Atari 2600 was designed to use an analog TV as the output device with 60 new frames appearing on the screen every second. To simplify the search space we impose a rule that one action is repeated over a fixed number of frames. This fixed number is called the *frame skip*. The standard frame skip used in [17] is 4. For this frame skip the agent makes a decision about the next move every $4 \cdot \frac{1}{60} = \frac{1}{15}$ of a second. Once the decision is made, then it remains unchanged during the next 4 frames.

Low frame skip allows the network to learn strategies based on a super-human reflex. High frame skip will limit the number of strategies, hence learning may be faster and more successful.

In the benchmark agent `nips`, trained with the frame skip 4, *all* 4 frames are used for training along with the sum of the rewards coming after them. This is dictated by the fact that due to hardware limitations, Atari games sometimes "blink", i.e. show some objects only every few frames. For example, in the game Space Invaders, if an enemy spaceship shoots in the direction of the player, then shots can be seen on the screen only every second frame and an agent who sees only the frames of the wrong parity would have no access to a critical part of the game information.

In the case of learning from memory we are not aware of any critical loses of information when intermediate RAM states are ignored. Hence in our models we only passed to the network the RAM state corresponding to the last frame corresponding to a given action[8].

The work [13] suggests that choosing the right frame skip can have a big influence on the performance of learning algorithms (see also [14]). Figure 6 and Table 4 show a significant improvement of the performance of the `just_ram` model in the case of Seaquest. Quite surprisingly, the variance of results appeared to be much lower for higher `FRAME SKIP`.

As noticed in [13], in the case of Breakout high frame skips, such as `FRAME SKIP = 30`, lead to a disastrous performance. Hence we tested only lower `FRAME SKIP` and for `FRAME SKIP = 8` we received results slightly weaker than those with `FRAME SKIP = 4`.

Table 4. Table summarizing test results for training which involves higher `FRAME SKIP` value. For Breakout `FRAME SKIP = 30` does not seem to be a suitable choice.

	Breakout	Seaquest
`just_ram` with `FRAME SKIP` 8 best	82.87	2064.44
`just_ram` with `FRAME SKIP` 30 best	–	2093.24
`big_ram` with `FRAME SKIP` 8 best	102.64	2284.44
`big_ram` with `FRAME SKIP` 30 best	–	2043.68

7   Mixing Screen and Memory

One of the hopes of future work is to integrate the information from the RAM and information from the screen in order to train an ultimate Atari 2600 agent. In this work we made some first steps towards this goal. We consider two mixed network architectures. The first one is `mixed_ram`, where we just concatenate the output of the last hidden layer of the convolutional network with the RAM input and then in the last layer apply a linear transformation without any following non-linearity.

---

[8] We also tried to pass all the RAM states as a $(128*\texttt{FRAME SKIP})$-dimensional vector, but this did not lead to an improved performance.

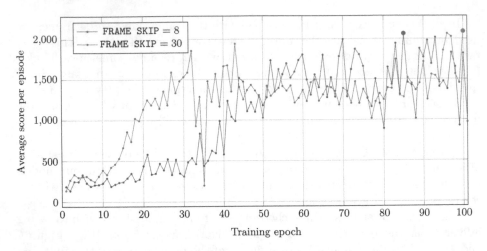

Fig. 6. Training results for Seaquest with FRAME SKIP = 8 and FRAME SKIP = 30 for the model just_ram.

Neural network 3. mixed_ram(outputDim)

Input: RAM,screen
Output: A vector of length outputDim

1 $conv1 \leftarrow Conv2DLayer(screen, rectify)$
2 $conv2 \leftarrow Conv2DLayer(conv1, rectify)$
3 $hidden \leftarrow DenseLayer(conv2, 256, rectify)$
4 $concat \leftarrow ConcatLayer(hidden, RAM)$
5 $output \leftarrow DenseLayer(concat, outputDim, no\ activation)$
6 **return** $output$

The other architecture is a deeper version of mixed_ram. We allow more dense layers which are applied in a more sophisticated way as described below.

Neural network 4. big_mixed_ram(outputDim)

Input: RAM,screen
Output: A vector of length outputDim

1 $conv1 \leftarrow Conv2DLayer(screen, rectify)$
2 $conv2 \leftarrow Conv2DLayer(conv1, rectify)$
3 $hidden1 \leftarrow DenseLayer(conv2, 256, rectify)$
4 $hidden2 \leftarrow DenseLayer(RAM, 128, rectify)$
5 $hidden3 \leftarrow DenseLayer(hidden2, 128, rectify)$
6 $concat \leftarrow ConcatLayer(hidden1, hidden3)$
7 $hidden4 \leftarrow DenseLayer(concat, 256, rectify)$
8 $output \leftarrow DenseLayer(hidden4, outputDim, no\ activation)$
9 **return** $output$

The obtained results presented in Table 5 are reasonable, but not particularly impressive. In particular we did not notice any improvement over the benchmark

Table 5. Table summarizing test results for methods involving information from the screen and from the memory.

	Breakout	Seaquest
`mixed_ram` best	143.67	488.57
`big_mixed_ram` best	67.56	1700

`nips` network, which is embedded into both mixed architectures. This suggests that in the `mixed_ram` and `big_mixed_ram` models the additional information from the memory is not used in a productive way.

8   RAM Visualization

We visualized the first layers of the neural networks in an attempt to understand how they work. Each column in Fig. 7 corresponds to one of 128 nodes in the first layer of the trained `big_ram` network and each row corresponds to one of 128 memory cells. The color of a cell in a given column describes whether the high value in this RAM cell negatively (blue) or positively (red) influence the activation level for that neuron. Figure 7 suggests that the RAM cells with numbers 95–105 in Breakout and 90–105 in Seaquest are important for the gameplay – the behavior of `big_ram` networks depend to the high extent on the state of these cells.

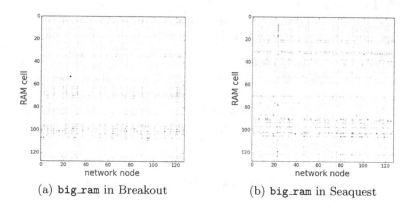

(a) `big_ram` in Breakout   (b) `big_ram` in Seaquest

Fig. 7. Visualization of the parameters of the first layer of the trained Q-networks. (Color figure online)

9   Conclusions

We trained a number of neural networks capable of playing Atari 2600 games: Bowling, Breakout and Seaquest. The novel aspect of this work is that the networks use information stored in the memory of the console. In all games the

RAM agents are on a par with the screen-only agent nips. The RAM agents trained using methods described in this work were unaware of more abstract features of the games, such as counters controlling amount of oxygen or the number of divers in Seaquest.

In the case of Seaquest, even a simple just_ram architecture with an appropriately chosen FRAME SKIP parameter as well as the big_ram agent with standard parameters, performs better than the benchmark nips agent. In the case of Breakout, the performance is below the screen-only agent nips, but still reasonable. In the case of Bowling methods presented in [17] as well as those in this paper are not very helpful – the agents play at a rudimentary level.

10   Future Work

10.1   Games with More *Refined* Logic

Since in the case of Seaquest the performance of RAM-only networks is quite good, a natural next target would be games such as Pacman or Space Invaders, which similarly to Seaquest offer interesting tactical challenges.

10.2   More Sophisticated Architecture and Better Hyperparameters

The recent papers [15,18,21,23] introduce more sophisticated ideas to improve deep Q-networks. We would like to see whether these improvements also apply to the RAM models.

It would be also interesting to tune hyperparameters in a way which would specifically address the needs of RAM-based neural networks. In particular we are interested in:

- better understanding what the deep Q-network learns about specific memory cells; can one identify critical cells in the memory?
- improving stability of learning and reducing variance and overfitting,
- more effective joining of information from the screen and from the memory,
- trying more complex, deeper architectures for RAM.

10.3   Recurrent Neural Networks and Patterns of Memory Usage

Reading the RAM state while running the deep Q-learning algorithm gives us an access to a practically unlimited stream of Atari 2600 memory states. We can use this stream to build a recurrent neural network which takes into account previous RAM states.

In our view it would also be interesting to train an *autoencoder*. It may help to identify RAM patterns used by Atari 2600 programmers and to find better initial parameters of the neural network [22].

10.4   Patterns as Finite State Machines

The work of Angluin [10] introduced the concept of learning the structure of a finite state machine through queries and counter-examples. A game for Atari 2600 can be identified with a finite state machine which takes as input the memory state and action and outputs another memory state. We are interested in devising a neural network which would learn the structure of this finite state machine. The successive layers of the network would learn about sub-automata responsible for specific memory cells and later layers would join the automata into an automaton which would act on the whole memory state.

Acknowledgements. This research was carried out with the support of grant GG63-11 awarded by the Interdisciplinary Centre for Mathematical and Computational Modelling (ICM) University of Warsaw. We would like to express our thanks to Marc G. Bellemare for suggesting this research topic.

A   Parameters

The list of hyperparameters and their descriptions. Most of the descriptions come from [18] (Table 6).

Table 6. Parameters

Hyperparameter	Value	Description
Minibatch size	32	Number of training cases over which each stochastic gradient descent (SGD) update is computed
Replay memory size	100 000	SGD updates are randomly sampled from this number of most recent frames
Phi length	4	The number of most recent frames experienced by the agent that are given as input to the Q network in case of the networks that accept screen as input
Update rule	rmsprop	Name of the algorithm optimizing the neural network's objective function
Learning rate	0.0002	The learning rate for rmsprop
Discount	0.95	Discount factor γ used in the Q-learning update. Measures how much less do we value our expectation of the value of the state in comparison to observed reward
Epsilon start	1.0	The probability (ε) of choosing a random action at the beginning of the training
Epsilon decay	1000000	Number of frames over which the ε is faded out to its final value
Epsilon min	0.1	The final value of ε, the probability of choosing a random action
Replay start size	100	The number of frames the learner does just the random actions to populate the replay memory

References

1. The Bowling Manual. https://atariage.com/manual_html_page.php?SoftwareID =879
2. Bowling (video game). https://en.wikipedia.org/wiki/Bowling_(video_game)
3. The Breakout Manual. https://atariage.com/manual_html_page.php?SoftwareID =889
4. Breakout (video game). https://en.wikipedia.org/wiki/Breakout_(video_game)
5. Lasagne - lightweight library to build and train neural networks in Theano. https:// github.com/lasagne/lasagne
6. Nathan Sprague's implementation of DQN. https://github.com/spragunr/deep_q_rl
7. The repository of our code. https://github.com/sygi/deep_q_rl
8. The Seaquest manual. https://atariage.com/manual_html_page.html?Software LabelID=424
9. Seaquest (video game). https://en.wikipedia.org/wiki/Seaquest_(video_game)
10. Angluin, D.: Learning regular sets from queries and counterexamples. Inf. Comput. **75**(2), 87–106 (1987)
11. Bellemare, M.G., Naddaf, Y., Veness, J., Bowling, M.: The arcade learning environment: an evaluation platform for general agents. J. Artif. Intell. Res. **47**, 253–279 (2013)
12. Bergstra, J., Breuleux, O., Bastien, F., Lamblin, P., Pascanu, R., Desjardins, G., Turian, J., Warde-Farley, D., Bengio, Y.: Theano: a CPU and GPU math expression compiler. In: Proceedings of the Python for Scientific Computing Conference (SciPy), Oral Presentation (2010)
13. Braylan, A., Hollenbeck, M., Meyerson, E., Miikkulainen, R.: Frame skip is a powerful parameter for learning to play Atari. In: AAAI-15 Workshop on Learning for General Competency in Video Games (2015)
14. Defazio, A., Graepel, T.: A comparison of learning algorithms on the Arcade learning environment. CoRR abs/1410.8620 (2014). http://arxiv.org/abs/1410.8620
15. Liang, Y., Machado, M.C., Talvitie, E., Bowling, M.: State of the art control of Atari games using shallow reinforcement learning. arXiv preprint arXiv:1512.01563 (2015)
16. Lipovetzky, N., Ramirez, M., Geffner, H.: Classical planning with simulators: results on the Atari video games. In: International Joint Conference on Artificial Intelligence (IJCAI), pp. 1610–1616 (2015)
17. Mnih, V., Kavukcuoglu, K., Silver, D., Graves, A., Antonoglou, I., Wierstra, D., Riedmiller, M.: Playing Atari with deep reinforcement learning. In: NIPS Deep Learning Workshop (2013)
18. Mnih, V., Kavukcuoglu, K., Silver, D., Rusu, A.A., Veness, J., Bellemare, M.G., Graves, A., Riedmiller, M., Fidjeland, A.K., Ostrovski, G., Petersen, S., Beattie, C., Sadik, A., Antonoglou, I., King, H., Kumaran, D., Wierstra, D., Legg, S., Hassabis, D.: Human-level control through deep reinforcement learning. Nature **518**(7540), 529–533 (2015)
19. Russell, S.J., Norvig, P.: Artificial Intelligence - A Modern Approach, 3 internat edn. Pearson Education, Englewood Cliffs (2010)
20. Srivastava, N., Hinton, G., Krizhevsky, A., Sutskever, I., Salakhutdinov, R.: Dropout: a simple way to prevent neural networks from overfitting. J. Mach. Learn. Res. **15**, 1929–1958 (2014)

21. Van Hasselt, H., Guez, A., Silver, D.: Deep reinforcement learning with double Q-learning. arXiv preprint arXiv:1509.06461 (2015)
22. Vincent, P., Larochelle, H., Bengio, Y., Manzagol, P.A.: Extracting and composing robust features with denoising autoencoders. In: Proceedings of the 25th International Conference on Machine Learning. ICML 2008, pp. 1096–1103. ACM, New York (2008)
23. Wang, Z., Schaul, T., Hessel, M., van Hasselt, H., Lanctot, M., de Freitas, N.: Dueling network architectures preprint arXiv:1511.06581 (2015)
24. Warde-Farley, D., Goodfellow, I.J., Courville, A., Bengio, Y.: An empirical analysis of dropout in piecewise linear networks. In: ICLR 2014 (2014)
25. Watkins, C.J.C.H., Dayan, P.: Technical note Q-learning. Mach. Learn. **8**, 279–292 (1992)

Clustering-Based Online Player Modeling

Jason M. Bindewald, Gilbert L. Peterson$^{(\boxtimes)}$, and Michael E. Miller

Air Force Institute of Technology, Wright-Patterson AFB, OH, USA
gilbert.peterson@afit.edu

Abstract. Being able to imitate individual players in a game can benefit game development by providing a means to create a variety of autonomous agents and aid understanding of which aspects of game states influence game-play. This paper presents a clustering and locally weighted regression method for modeling and imitating individual players. The algorithm first learns a generic player cluster model that is updated online to capture an individual's game-play tendencies. The models can then be used to play the game or for analysis to identify how different players react to separate aspects of game states. The method is demonstrated on a tablet-based trajectory generation game called *Space Navigator*.

1 Introduction

Automating game-play in a human-like manner is one goal in intelligent gaming research, with applications such as a gaming version of the Turing Test [14] and human-like game avatars [6]. When we move from playing a game generically to playing like a specific individual, the dynamics of the problem change [10]. In complex dynamic environments, it can be difficult to differentiate individual players, because the insights exploited in imitating 'human-like' game-play can become less useful in imitating the idiosyncrasies that differentiate specific individuals' game-play. By learning how to imitate individual player behaviors, we can model more believable opponents [6] and understand what demarcates individual players, which allows a game designer to build robust game personalization [18].

The *Space Navigator* environment provides a test-bed for player modeling in routing tasks, and allows us to see how different game states affect disparate individuals' performance of a routing task. The routing task is a sub-task of several more complex task environments, such as real-time strategy games or air traffic control tasks. Since there is only one action a player needs to take:

M.E. Miller—The views expressed in this document are those of the author and do not reflect the official policy or position of the United States Air Force, the United States Department of Defense, or the United States Government. This work was supported in part through the Air Force Office of Scientific Research, Computational Cognition & Robust Decision Making Program (FA9550), James Lawton Program Manager.

© Springer International Publishing AG (outside the US) 2017
T. Cazenave et al. (Eds.): CGW 2016/GIGA 2016, CCIS 705, pp. 86–100, 2017.
DOI: 10.1007/978-3-319-57969-6_7

draw a trajectory, it is easy for players to understand. However, *Space Navigator*, with its built in dynamism, is complex enough that it is not simple to generate a single 'best input' to any given game state. The dynamism also means that replaying an individual's past play is not possible.

Specifically, we use individual player modeling to enable a trajectory generator that acts in response to game states in a manner that is similar to what a specific individual would have done in the same situation. Individualized response generation enables better automated agents within games that can imitate individual players for reasons such as creating "stand-in" opponents or honing strategy changes by competing against oneself. In addition, the player models can be used by designers to identify where the better players place emphasis and how they play, which can be used to balance gameplay or create meaningful tutorials.

This paper contributes a player modeling paradigm that enables an automated agent to perform response actions in a game that are similar to those that an individual player would have performed. The paradigm is broken into three steps: (1) a cluster-based generic player model is created offline, (2) individual player models hone the generic model online as players interact with the game, and (3) responses to game situations utilize the individual player models to imitate the responses players would have given in similar situations. The resulting player models can point game designers toward the areas of a game state that affect individual behavior in a routing task in more or less significant ways.

The remainder of the paper proceeds as follows. Section 2 reviews related work. Section 3 introduces the *Space Navigator* trajectory routing game. Section 4 presents the online individual player modeling paradigm and the model is then applied to the environment in Sect. 5. Section 6 gives experimental results showing the individual player modeling system's improvements over a generic modeling method for creating trajectories similar to individual users. Section 7 summarizes the findings presented and proposes potential future work.

2   Related Work

Player models can be grouped across four independent facets [15]: domain, purpose, scope, and source. The domain of a player model is either game actions or human reactions. Purpose describes the end for which the player model is implemented: generative player models aim to generate actual data in the environment in place of a human or computer player, while descriptive player models aim to convey information about a player to a human. Scope describes the level of player(s) the model represents: individual (one), class (a group of more than one), universal (all), and hypothetical (other). The source of a player model can be one of four categories: induced - objective measures of actions in a game; interpreted - subjective mappings of actions to a pre-defined category; analytic - theoretical mappings based on the game's design; and synthetic - based on non-measurable influence outside game context. As an example classification, the player model created in [16] for race track generation models individual player tendencies and preferences (Individual), objectively measures actions in the game (Induced), creates tracks in the actual environment (Generative), and arises from game-play data (Game Action). The player model created here furthers this work by updating the player model online.

One specific area of related work in player modeling involves player decision modeling. Player decision modeling [8] aims to reproduce the decisions that players make in an environment or game. These models don't necessarily care why a given decision was made as long as the decisions can be closely reproduced. Utilizing player decision modeling, procedural personas [7,11] create simple agents that can act as play testers. By capturing the manner in which a given player or set of players makes decisions when faced with specific game states, the personas can help with low-level design decisions.

Past work has used Case-Based Reasoning (CBR) [5] and Learning from Demonstration (LfD) [1] to translate insights gained through player modeling into responses within an environment. The nearest neighbor principle, maintaining that instances of a problem that are a shorter distance apart more closely resemble each other than do instances that are a further distance apart, is used to find relevant past experiences in LfD tasks such as a robot intercepting a ball [1], CBR tasks such as a RoboCup soccer-playing agent [5], or tasks integrating both LfD and CBR such as in real time strategy games [13]. When searching through large databases of past experiences approximate nearest neighbors searches, such as Fast Library for Approximate Nearest Neighbors (FLANN [12]), have proven useful in approximating nearest neighbor searches while maintaining lower order computation times in large search spaces.

3   Application Environment

Space Navigator [2,3] is a tablet computer game similar to *Flight Control* [4] and *Contrails* [9]. Figure 1 shows a screen capture from the game and identifies several key objects within the game. Spaceships appear at set intervals from the screen edges. The player directs each spaceship to its destination planet (designated by similar color) by drawing a line on the game screen using his or her finger. Points accumulate when a ship encounters its destination planet or bonuses that randomly appear throughout the play area. Points decrement when spaceships collide and when a spaceship traverses one of several "no-fly zones" (NFZs) that move throughout the play area at a set time interval. The game ends after five minutes.

4   Methodology

The player modeling paradigm shown in Fig. 2 begins with a cluster-based generic player model created offline (area 1). The generic player model is updated online to adapt to changing player habits and quickly differentiate between players (area 2). Then the online player modeler creates responses to game states that are similar to those that an individual player would have given in response to similar states (area 3).

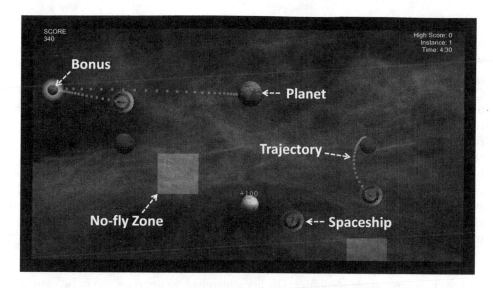

Fig. 1. A *Space Navigator* screen capture highlighting important game objects. (Color figure online)

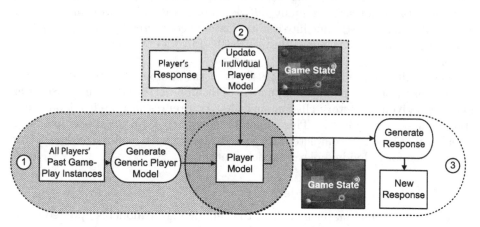

Fig. 2. An online updating individual player modeling paradigm.

State and Response Clustering. Clustering reduces the state-response pairs into a set of representative clusters, dramatically reducing the representation size of a player model. Ward agglomerative clustering [17] provides a baseline for the player modeling method and was proven effective for clustering in trajectory creation game environments in [3,9]. The clustering implemented here takes each game-play instance, containing a state and its associated response, and assigns it to both a state cluster and a response cluster. The number of clusters is a choice left to the practitioner, accounting for the specific environment and resource constraints. A state-response instance mapping from a given state cluster to a given response cluster demonstrates a proclivity for a player to react with a class of maneuver in a specific type of game situation. By determining the frequency of mappings, common situational responses and outlier actions emerge.

Cluster Outlier Pruning. If a state has only been seen in one instance by one player, that state is unlikely to provide much benefit in predicting future responses. After state and response clustering, clusters with outlier responses are removed first by removing all instances assigned to the least populated response clusters. The cutoff threshold for determining which instances to remove could be either a minimum response cluster size or a percentage of response clusters to remove. For example, due to the distribution of cluster sizes in the *Space Navigator* database we removed instances falling in the bottom 25% of all response clusters according to cluster size (setting a cutoff threshold relies on knowledge of the environment and underlying dataset distribution, and is an area for future work).

Similarly, a response given by only one player in one instance is unlikely to reoccur in future player responses. Outlier state clusters are removed in two ways. First, instances that fall in the bottom 25% of all state clusters according to cluster size are removed, eliminating response clusters that are rare overall. However, removing states not seen by many *different* players is also important. Pruning also removes instances falling into a state cluster encountered by a minimal subset of players, eliminating response clusters reached by an extremely small subset of players.

The resulting player model, $\mathbf{P}_{x,y}$ is the (x = the number of state clusters) $\times$ (y = the number of response clusters) matrix of likelihoods that a given state cluster maps to a given trajectory cluster, such that $\mathbf{p}_{i,j}$ represents the likelihood that state s_i maps to response r_i. This model is created across all game-play instances after cluster pruning is complete. This generic player model, created off-line, forms the baseline for individual player model creation.

4.1   Individual Player Models

For online individual player modeling, the generic player model is updated as an individual plays the game (shaded area 2 of Fig. 2). Over time, the updates shape a player model of an individual player's game-play tendencies. The individual player update trains quickly by weighting learning according to state-response cluster scores.

Algorithm 1 is the online algorithm for learning individual player models. The algorithm begins with the generic player model $\mathbf{P}$. Once a player submits a response in the game environment, the current game state and the response are given as inputs. The algorithm finds the closest state (S_{close}) and response (R_{close}) clusters, and the player model is updated at the intersection of S_{close} and R_{close} by δ_{close}. Then the player model is normalized across all the R values for S_{close} so that the values sum to 1.

There are certain states that provide more information than others. Weighting the increment values for a given state-trajectory pair aids quick learning of

Algorithm 1. Individual player model online update algorithm.

1: **inputs**: $\mathbf{P} = x \times y$ generic player model; $\langle s_{in}, r_{in} \rangle$ = a state-response pair; $\mathbf{M} =$
        $\{\langle S_1, R_1 \rangle, \langle S_1, R_2 \rangle, \cdots, \langle S_x, R_y \rangle\}$, all cluster mappings
2: S_{close} = the closest state cluster to state s_{in}
3: $\delta_{close} = q \cdot (\delta_{cp} + \delta_{cmv} + \delta_{pma})$, S_{close}'s update increment weight
4: R_{close} = the closest response cluster to response r_{in}
5: $\mathbf{p}(S_{close}, R_{close}) = \mathbf{p}(S_{close}, R_{close}) + \delta_{close}$
6: **for** $\mathbf{p}(S_{close}, i)$ where $i = 1 \rightarrow y$ **do**
7:        $\mathbf{p}(S_{close}, i) = \mathbf{p}(S_{close}, i) / (1 + \delta_{close})$
8: **end for**

player idiosyncrasies. Traits gleaned from the clustered data help determine which
state clusters should create larger learning increments, and which states provide
minimal information beyond the generic player model. Three traits comprise the
update increment, δ. As shown in Algorithm 1, Line 3 these include: cluster population, cluster mapping variance, and previous modeling utility.

Cluster Population: When attempting to learn game-play habits quickly,
knowing the expected responses of a player to common game states is important. Weighting δ according to the size of a state cluster in comparison to that of
the other state clusters across the entire game-play dataset emphasizes increased
learning from common states for an individual player model. States that fall into
larger clusters can provide better information for quickly learning to differentiate individual player game-play. To calculate the cluster population trait, all state
cluster sizes are calculated and any state cluster with a population above a selected
population threshold is given a cluster population trait weight of $\delta_{cp} = 1$ and all
other state clusters receive a weight of $\delta_{cp} = 0$.

Cluster Mapping Variance: When mapping state clusters to response clusters,
some state clusters will consistently map to a specific response cluster across all
players. Other state clusters will consistently map to several response clusters
across all players. Very little about a player's game-play tendencies is learned from
these two types of state clusters. However, state clusters that map to relatively few
clusters per player (intra-player cluster variance), while still varying largely across
all players (inter-player cluster variance) can help quickly differentiate players.
The state cluster mapping variance ratio is the total number of response clusters
to which a state cluster maps across all players divided by the number of response
clusters to which the average player maps, essentially the ratio of inter-player cluster variance to the intra-player cluster variance. The cluster mapping variance
trait weight, δ_{cmv}, is set according to a cluster variance ratio threshold. All state
clusters with a variance ratio above the threshold receive a weight of $\delta_{cmv} = 1$ and
all others receive a weight of $\delta_{cmv} = 0$.

Previous Modeling Utility: The last trait involves running Algorithm 1 on the
existing game-play data. Running the individual player update model on previous game-play data provides insight into how the model works in the actual game
environment. First, Algorithm 1 runs with $\delta = 1$ for all state clusters, training
the player model on some subset of a player's game-play data (training set). Then

it iterates through the remaining game-play instances (test set) and generates a response to each presented state, using both the individual player model and the generic player model. This iteration includes each individual player in the game-play dataset. For each test set state, the response most similar to the player's actual response is determined. Each time the individual player model is closer than the generic player model to the actual player response, tally a 'win' for the given state cluster and a 'loss' otherwise. The ratio of wins to losses for each state cluster makes up the previous modeling utility trait. The previous modeling utility trait weight, δ_{pma}, is set according to a previous modeling utility threshold. All state clusters with a previous modeling utility above the threshold receive a weight of $\delta_{pma} = 1$ and all others receive a weight of $\delta_{pma} = 0$.

Calculating δ: When Algorithm 1 runs, δ is set to the sum of all trait weights for the given state cluster multiplied by some value q which is an experimental update increment set by the player. Line 3 shows how δ is calculated as a sum of the previously discussed trait weights.

4.2   Generate Response

Since response generation is environment specific, this section demonstrates the response generation section shown in area 3 of Fig. 2 for a trajectory generation task. The resulting trajectory generator creates trajectories that imitate a specific player's game-play, using the cluster weights in $\mathbf{P}$ from either a generic or learned player model.

The trajectory response generation algorithm takes as input: the number of trajectories to weight and combine for each response (k), the number of state and trajectory clusters (x and y respectively), the re-sampled trajectory size (μ), a new state (s_{new}), a player model ($\mathbf{P}$), and the set of all state-trajectory cluster mappings ($\mathbf{M}$). Line 2 begins by creating an empty trajectory of length μ which will hold the trajectory generator's response to s_{new}. Line 3 then finds the state cluster (S_{close}) to which s_{new} maps. $\mathbf{P}_{close}$, created in Line 4, contains a set of likelihoods. $\mathbf{P}_{close}$ holds the likelihoods of the k most likely trajectory clusters to which state cluster S_{close} maps.

The loop at Line 5 then builds the trajectory response to s_{new}. Humans tend to think in terms of 'full maneuvers' when generating trajectories–specifically for very quick trajectory generation tasks such as trajectory creation games [9]–rather than creating trajectories one point at a time. Therefore, the *Space Navigator* trajectory response generator creates full maneuver trajectories. Line 7 finds the instance assigned to both state cluster S_{close} and trajectory cluster T_i with the state closest to s_{new}. The response to this state is then weighted according to the likelihoods in $\mathbf{P}$. The loop in Line 9, then combines the k trajectories using a weighted average for each of the μ points of the trajectory. The weighted average trajectory points are normalized across the k weights used for the trajectory combination in Line 13 and returned as the response to state s_{new} according to the player model $\mathbf{P}$.

Algorithm 2. Trajectory response generation algorithm.

1: **inputs**: k = the number of trajectories to combine; x = the number of state clusters; y = the number of trajectory clusters; μ = the re-sampled trajectory size; s_{new} = a state we have not seen before; $\mathbf{P}$ = an $x \times y$ player model; $\mathbf{M} = \{\langle S_1, T_1 \rangle, \langle S_1, T_2 \rangle, \cdots, \langle S_x, T_y \rangle\}$, all state-trajectory cluster mappings

2: **initialize**: $t_{new}(\mu) \leftarrow$ an empty trajectory of μ points

3: S_{close} = the closest state cluster to state s_{new}

4: $\mathbf{P}_{close} = \max\limits_{k} \left[\mathbf{P}_{S_{close}, (z | \forall z \in 1,\ldots,y)} \right]$

5: **for each** $P_{close,i} \in \mathbf{P}_{close}$ **do**

6:     T_i = the trajectory cluster associated with $P_{close,i}$

7:     $s_{close,i} \leftarrow$ state closest to s_{new} in $\langle S_{close}, T_i \rangle$

8:     $t_{close,i} \leftarrow$ the response trajectory to $s_{close,i}$

9:     **for** $\nu = 1 \rightarrow \mu$ **do**

10:         $t_{new}(\nu) = t_{new}(\nu) + t_{close,i}(\nu) \cdot P_{close,i}$

11:     **end for**

12: **end for**

13: **return** $t_{new} = t_{new} / \sum\limits_{i=1}^{k} P_{close,i}$

5 Environment Considerations

This section demonstrates how the player modeling paradigm can be applied to generating trajectory responses in *Space Navigator*. First, an initial data capture experiment is outlined. Then, solutions are presented to two environment specific challenges: developing a state representation and comparing disparate trajectories.

5.1 Initial Data Capture Experiment

An initial experiment captured a corpus of game-play data for further comparison and benchmarking of human game-play [3]. Data was collected from 32 participants playing 16 five-minute instances of *Space Navigator*. The instances represented four difficulty combinations, with two specific settings changing: (1) the number of NFZs and (2) the rate at which new ships appear. The environment captures data associated with the game state when the player draws a trajectory, including: time stamp, current score, ship spawn rate, NFZ move rate, bonus spawn interval, bonus info (number, location, and lifespan of each), NFZ info (number, location, and lifespan of each), other ship info (number, ship ID number, location, orientation, trajectory points, and lifespan of each), destination planet location, selected ship info (current ship's location, ship ID number, orientation, lifespan, and time to draw the trajectory), and selected ship's trajectory points. The final collected dataset consists of 63,030 instances.

5.2 State Representation

Space Navigator states are dynamic both in number and location of objects. The resulting infinite number of configurations makes individual state identification difficult. To reduce feature vector size, the state representation contains only the elements of a state that directly affect a player's score (other ships, bonuses, and NFZs) scaled to a uniform size, along with a feature indicating the relative length of the spaceship's original distance from its destination. Algorithm 3 describes the state-space feature vector creation process.

Algorithm 3. State-space feature vector creation algorithm.

1: **input**: L = the straight-line trajectory from spaceship to destination planet.
2: **initialize**: $\eta \in [0.0 \cdots 1.0]$ = weighting variable; s = empty array (length 19);
$\quad\quad zoneCount = 1$
3: Translate all objects equally s.t. the selected spaceship is located at the origin.
4: Rotate all objects in state-space s.t. L lies along the X-axis.
5: Scale state-space s.t. L lies along the line segment from $(0,0)$ to $(1,0)$.
6: **for** each object type $\vartheta \in (OtherShip, Bonus, NFZ)$ **do**
7: $\quad$ **for** each zone $z = 1 \rightarrow 6$ **do**
8: $\quad\quad zoneCount = zoneCount + 1$
9: $\quad\quad$ **for** each object o of type ϑ in zone z **do**
10: $\quad\quad\quad d_o$ = the shortest distance of o from L
11: $\quad\quad\quad w_o = e^{-(\eta \cdot d_o)^2}$ $\quad\quad\quad\quad\quad\quad\quad$ ▷ Gaussian weight function
12: $\quad\quad\quad s\,[zoneCount] = s\,[zoneCount] + w_o$
13: $\quad\quad$ **end for**
14: $\quad$ **end for**
15: **end for**
16: $s\,[19]$ = the non-transformed straight-line trajectory length
17: **return** s, normalized between $[0,1]$

The algorithm first transforms the state-space features against a straight-line trajectory frame in Line 1. Lines 3–5 transform the state-space along the straight-line trajectory such that disparate trajectories can be compared in the state-space. The loop at Line 6 accounts for different element types and the loop at Line 7 divides the state-space into six zones as shown in Fig. 3. This effectively divides the state-space into left and right regions, each with three zones with relation to the spaceship's straight-line path: behind the spaceship, along the path, and beyond the destination.

To compare disparate numbers of objects, the loop beginning in Line 9 uses a weighting method similar to that used in [5], collecting a weight score (s) for each object within the zone. This weight score is calculated using a Gaussian weighting function based on the minimum distance an object is from the straight-line trajectory. Figure 3 shows the transformation of the state into a feature vector using Algorithm 3. The state-space is transformed in relation to the straight-line trajectory, and a value is assigned to each "entity type + zone" pair accordingly. For example, Zone 1 has a bonus value of 0.11 and other ship and NFZ values of 0.00, since it only contains one bonus. Lastly, the straight-line trajectory distance is captured. This accounts for the different tactics used when ships are at different distances from their destination. The resulting state representation values are normalized between zero and one.

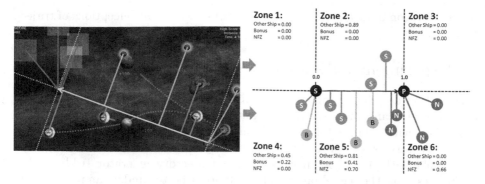

Fig. 3. The six zones surrounding the straight line trajectory in a *Space Navigator* state representation and the state representation calculated with Algorithm 3.

5.3   Trajectory Comparison

Trajectory generation requires a method to compare disparate trajectories. Trajectory re-sampling addresses the fact that trajectories generated within *Space Navigator* vary in composition, containing differing numbers of points and point locations. Re-sampling begins by keeping the same start and end points, and iterates until the re-sampled trajectory is filled. The process first finds the proportional relative position (p_m) of a point. The proportional relative position indicates where the i-th point would have fallen in the original trajectory and may fall somewhere between two points. The proportional distance (d_m) that p_m falls from the previous point in the old trajectory (p_0) is the relative distance that the i-th re-sampled point falls from the previous point. To compare trajectories, the target number of points is set to 50 (approximately the mean trajectory length in the initial data capture) for re-sampling all the trajectories.

Re-sampling has two advantages: the re-sampling process remains the same for both trajectories that are too long and too short and maintains the distribution of points along the trajectory. A long or short distance between two consecutive points, remains in the re-sampled trajectory. This ensures that trajectories drawn quickly or slowly maintain those sampling characteristics despite the fact that the draw rate influences the number of points in the trajectory. Since feature vector creation geometrically transforms a state, the trajectories generated in response to the state are transformed in the same manner, ensuring the state-space and trajectory response are positioned in the same state space.

To ensure the trajectories generated in *Space Navigator* are similar to those of an individual player, a distance measure captures the objective elements of trajectory similarity. The Euclidean trajectory distance treats every trajectory of i (x, y) points, as a $2i$-dimensional point in Euclidean space where each x and each y value in the trajectory represents a dimension. The distance between two trajectories is the simple Euclidean distance between the two $2i$-dimensional point representations of the trajectories. A 35-participant human-subject study confirmed that Euclidean trajectory distance not only distinguished between trajectories computationally, but also according to human conceptions of trajectory similarity.

6 Experiment and Results

This section describes testing of the online individual player modeling trajectory generator and presents insights gained from the experiment. The results show that, with a limited amount of training data, the individual player modeling trajectory generator is able to create trajectories more similar to those of a given player than a generic player-modeling trajectory generator. Additionally, the results show the model provides insights for a better understanding of what separates different players' game-play via comparison to the generic player model.

6.1 Experiment Settings

The experiment compares trajectories created with the generic player model, the individual player model, and a generator that always draws a straight line between the spaceship and its destination planet. The first five games are set aside as a training dataset and the next eleven games as a testing dataset. Five training games (equivalent to 25 min of play) was chosen as a benchmark for learning an individual player model to force the system to quickly pull insights that would manifest in later game-play. For each of 32 players, the individual player model is trained on the five-game training dataset using Algorithm 1 with the trait score weights. Next, each state in the given player's testing set is presented to all three trajectory generators and the difference between the generated and actual trajectories recorded. Experimental values for the individual player model are set as follows: update increment (q) = 0.01, cluster population threshold = 240, cluster mapping variance threshold = 17.0, and previous modeling utility threshold = 3.0.

The three learning thresholds specific to *Space Navigator* are: (1) state cluster population threshold = 240 (set at a value of one standard deviation over the mean cluster size), forty of 500 state clusters received a cluster population weight of $\delta_{cp} = 1$; (2) cluster mapping variance ratio threshold = 17, with 461 of 500 state clusters receiving a cluster variance weight of $\delta_{cmv} = 1$; and (3) previous modeling utility threshold = 3, with 442 of 500 clusters receiving a previous modeling utility score of $\delta_{pma} = 1$.

To account for the indistinguishability of shorter trajectories, results were removed for state-trajectory pairs with straight-line trajectory length less than approximately 3.5 cm on tablets with 29.5 cm screens used for experiments. This distance was chosen as it represents the trajectory length at which an accuracy one standard deviation below the mean was reached.

6.2 Individual Player Modeling Results

Testing of the game-play databases shows that the trajectories generated using the individual player model significantly improved individual player imitation results

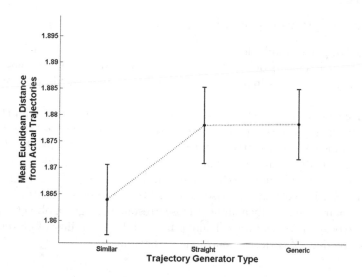

Fig. 4. Euclidean trajectory distance between generated trajectories and actual trajectory responses.

when compared to those generated by the generic player model and the straight line trajectory generator. Table 1 and Fig. 4 show results comparing trajectories generated using each database with the actual trajectory provided by the player, showing the mean Euclidean trajectory distance and standard error of the mean across all 32 players and instances.

Table 1. Mean and standard error of the Euclidean trajectory distances (in *SpaceNavigator* environment meters).

Database	Mean dist	Std err
Individual player model	1.8640	±0.0063
Straight line generator	1.8781	±0.0069
Generic player model	1.8784	±0.0063

The individual player model generator provides an improvement over the other models. The mean Euclidean trajectory distance of 1.8640 provides a statistically significant improvement over the straight line and generic player models, as standard error across all instances from all 32 players does not overlap with the latter two player models. The similar player model improves the generic databases accuracy by learning more from a selected subset of presented states to ensure that the player model more accurately generates similar trajectories.

6.3   Individual Player Model Insight Generation

The changes in player model learning value for each element of a state representation show which aspects of the state influence game-play. This enables a better understanding of what distinguishes individual game-play within the game environment.

Table 2 shows the results of a Pearson's linear correlation between the mean learning value change of each state cluster across all 32 players and the state representation values of the associated state cluster centroids. The results show that there is a statistically significant negative correlation between the mean learning value changes and all of the zones, but some changes are much larger than others. The overall negative correlation arises among object/zone pairs intuitively: high object/zone pair scores imply a large or close presence of a given object type, constraining the possible trajectories. There is more differentiability of player actions when more freedom of trajectory movement is available.

Table 2. Correlation of each state representation value with the mean change in associated state cluster learning values in player models

Zone	Pearson's r	p-value	Pearson's r	p-value	Pearson's r	p-value
	Other Ships		*Bonuses*		*NFZs*	
1	-0.1227	0.0060	-0.1569	0.0004	-0.1002	0.0251
2	-0.3911	0.0000	-0.3552	0.0000	-0.2749	0.0000
3	-0.1616	0.0003	-0.2212	0.0000	-0.1184	0.0080
4	-0.1465	0.0010	-0.1662	0.0002	-0.1159	0.0095
5	-0.4244	0.0000	-0.3693	0.0000	-0.2398	0.0000
6	-0.1903	0.0000	-0.2056	0.0000	-0.1040	0.0200
Dist	-0.6434	0.0000				

With the ship-to-planet distance feature, longer distances correlate to smaller learning value changes among player models, with the strongest correlation of all features: r of -0.6434 and p-value <0.0001. Possible explanations for this behavior include: (1) players are more constrained over long distances, (2) as distances get longer, the variance in the way an individual player draws trajectories in similar situations increases, (3) shorter distances capture consistent tendencies that carry along to distinguish individual game-play over time.

Another aspect that Table 2 shows is the importance of the middle zones in comparison to the 'before' and 'after' zones. Figure 5 illustrates this point graphically. The r values show that the middle two zones provide a larger influence on the amount of change in the learning values. For example, in Fig. 5a the r values for zones two and five are more than double those of any other zone. This idea is somewhat intuitive as this is the area that the ship will traverse, providing the most likely cause for interaction with objects of any given type.

The results also provide insight into the relative value that players place on certain types of objects. For example, determining the correlation coefficients of different Object/Zone Pairs can show that No Fly Zones in the middle two zones provide a significantly smaller influence on learning value changes than other ships do in the same zones.

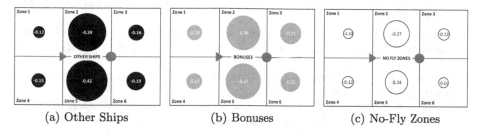

(a) Other Ships   (b) Bonuses   (c) No-Fly Zones

Fig. 5. Graphical representation of the correlation coefficient for each Object Type/Zone score with the mean change in learning values in player models.

Three examples of how player modeling insights can be used in game applications involve training, game design, and player automation. Player models can be used to find places where specific users who are doing really well properly value certain actions over others. Proper valuations can then be communicated to players during training within the environment. Another example is that, we can use the player modeling insights to design point structures to more closely align with the way players perceive the value of different object types. Lastly, modeling a specific player enables the designer to incorporate an automated player to play like a specific expert or current user within the game.

7   Conclusions and Future Work

The online individual player modeling paradigm presented in this paper is able to generate trajectories similar to those of a specific *Space Navigator* player. The system is able to operate online without needing to perform time-consuming offline calculations to update individual player models. Additionally, the gains in individual player imitation are found in a relatively small number of games (five games, totaling 25 min). The player models developed to imitate players also allow for a better understanding of what traits of a given state provide understanding of player differences which occur for different states.

This work provides opportunities for several areas of future work. Further studies will research the effects of using the trajectory generator to act as an automated aid for players interacting with the *Space Navigator* game. Additionally, further analysis of the player modeling methods could yield further insights into how much differentiation of individual players can be gained over different amounts of time. Moreover, imitating individual players could provide helpful insights in determining how experts play *Space Navigator* to aid in experiments to learn how to improve player training.

References

1. Argall, B., Browning, B., Veloso, M.: Learning by demonstration with critique from a human teacher. In: Proceedings of the ACM/IEEE International Conference on Human-Robot Interaction, pp. 57–64. ACM (2007)
2. Bindewald, J.M., Miller, M.E., Peterson, G.L.: A function-to-task process model for adaptive automation system design. Int. J. Hum. Comput. Stud. **72**(12), 822–834 (2014)
3. Bindewald, J.M., Peterson, G.L., Miller, M.E.: Trajectory generation with player modeling. In: Barbosa, D., Milios, E. (eds.) CANADIAN AI 2015. LNCS (LNAI), vol. 9091, pp. 42–49. Springer, Cham (2015). doi:10.1007/978-3-319-18356-5_4
4. Firemint Party Ltd.: Flight control, December 2011. https://itunes.apple.com/us/app/flight-control/id306220440?mt=8
5. Floyd, M.W., Esfandiari, B., Lam, K.: A case-based reasoning approach to imitating robocup players. In: FLAIRS Conference, pp. 251–256 (2008)
6. Gamez, D., Fountas, Z., Fidjeland, A.K.: A neurally controlled computer game avatar with human like behavior. IEEE Trans. Comput. Intell. AI Games **5**(1), 1–14 (2013)
7. Holmgård, C., Liapis, A., Togelius, J., Yannakakis, G.N.: Evolving personas for player decision modeling. In: 2014 IEEE Conference on Computational Intelligence and Games (CIG), pp. 1–8. IEEE (2014)
8. Holmgård, C., Liapis, A., Togelius, J., Yannakakis, G.N.: Generative agents for player decision modeling in games. Foundations of Digital Games (2014)
9. Huang, V., Huang, H., Thatipamala, S., Tomlin, C.J.: Contrails: crowd-sourced learning of human models in an aircraft landing game. In: Proceedings of the AIAA GNC Conference (2013)
10. Kemmerling, M., Ackermann, N., Preuss, M.: Making Diplomacy bots individual. In: Hingston, P. (ed.) Believable Bots, pp. 265–288. Springer, Heidelberg (2012)
11. Liapis, A., Holmgård, C., Yannakakis, G.N., Togelius, J.: Procedural personas as critics for dungeon generation. In: Mora, A.M., Squillero, G. (eds.) EvoApplications 2015. LNCS, vol. 9028, pp. 331–343. Springer, Cham (2015). doi:10.1007/978-3-319-16549-3_27
12. Muja, M., Lowe, D.G.: Fast approximate nearest neighbors with automatic algorithm configuration. VISAPP (1), 2 (2009)
13. Ontanón, S., Synnaeve, G., Uriarte, A., Richoux, F., Churchill, D., Preuss, M.: A survey of real-time strategy game AI research and competition in StarCraft. Comput. Intell. AI Games **5**(4), 293–311 (2013)
14. Schrum, J., Karpov, I.V., Miikkulainen, R.: Human-like combat behaviour via multiobjective neuroevolution. In: Hingston, P. (ed.) Believable Bots, pp. 119–150. Springer, Heidelberg (2012)
15. Smith, A.M., Lewis, C., Hullet, K., Sullivan, A.: An inclusive view of player modeling. In: The 6th International Conference on Foundations of Digital Games, pp. 301–303. ACM (2011)
16. Togelius, J., De Nardi, R., Lucas, S.M.: Making racing fun through player modeling and track evolution. In: Proceedings Optimizing Player Satisfaction in Computer and Physical Games, p. 61 (2006)
17. Ward, J.H.: Hierarchical grouping to optimize an objective function. J. Am. Stat. Assoc. **58**(301), 236–244 (1963)
18. Yu, H., Riedl, M.O.: Personalized interactive narratives via sequential recommendation of plot points. IEEE Trans. Comput. Intell. AI Games **6**(2), 174–187 (2014)

AI Wolf Contest — Development of Game AI Using Collective Intelligence —

Fujio Toriumi[1(✉)], Hirotaka Osawa[2], Michimasa Inaba[3], Daisuke Katagami[4], Kosuke Shinoda[5], and Hitoshi Matsubara[6]

[1] The University of Tokyo, 7-3-1 Hongo, Bunkyo-ku, Tokyo, Japan
tori@sys.t.u-tokyo.ac.jp
[2] University of Tsukuba, Tsukuba, Japan
[3] Hiroshima City University, Hiroshima, Japan
[4] Tokyo Polytechnic University, Tokyo, Japan
[5] The University of Electro-Communications, Chofu, Japan
[6] Future University Hakodate, Hakodate, Japan

Abstract. In this study, we specify the design of an artificial intelligence (AI) player for a communication game called "Are You a Werewolf?" (AI Wolf). We present the Werewolf game as a standard game problem in the AI field. It is similar to game problems such as Chess, Shogi, Go, and Poker. The Werewolf game is a communication game that requires several AI technologies such as multi-agent coordination, intentional reading, and understanding of the theory of mind. Analyzing and solving the Werewolf game as a standard problem will provide useful results for our research field and its applications. Similar to the RoboCup project, the goal of this project is to determine new themes while creating a communicative AI player that can play the Werewolf game with humans. As an initial step, we designed a platform to develop a game-playing AI for a competition. First, we discuss the essential factors in Werewolf with reference to other studies. We then develop a platform for an AI game competition that uses simplified rules to support the development of AIs that can play Werewolf. The paper reports the process and analysis of the results of the competition.

1 Introduction

The development of an artificial intelligence (AI) player that can play a game with a human has been one of the main benchmarks in the AI field for researching intelligence and its requirements. In the field of complete-information games, such as Chess or Shogi, AIs have already defeated top human players. In 2016, an AI defeated a human in the last-remaining complete-information game, Go [1]. In the field of incomplete games, Texas Hold'em Poker is a conventional game that organizes competitions in the AI field [2]. In 2015, Bowling et al. solved the problem of two-player-limited Poker [3]. Further, action video games are starting to be used for evaluating AI in real-time situations [4].

Compared to previous game challenges, communication or communicative intelligence, which is commonly used in board and card games, has not been attempted. When users play board and card games, they also converse with other players. Furthermore,

T. Cazenave et al. (Eds.): CGW 2016/GIGA 2016, CCIS 705, pp. 101–115, 2017.
DOI: 10.1007/978-3-319-57969-6_8

some games are actually conducted through conversations, and these are referred to as communication games. Relatively few studies focus on the application of AI in such communication games.

"Are You a Werewolf?" is one of the most popular communication games. The cover story of the "Werewolf" game (also known as "Mafia") is as follows. "It's a story about a village. Werewolves have arrived who can change into and eat humans. The werewolves have the same form as humans during the day, and attack the villagers one-by-one every night. Fear, uncertainty, and doubt towards the werewolves begin to grow. The villagers decide that they must execute those who are suspected of being werewolves, one by one…"

The winner of the Werewolf game is decided solely through discussions. Consequently, game players must use their cognitive faculties to the full. In contrast to a perfect-information game, players hide considerable information. Every player attempts to determine the hidden information by using other players' conversations and behaviors, while trying to hide his/her own information to accomplish the objective. The game highlights various problems that have not been addressed adequately in the area of AI, such as an asymmetric diversity of player information, persuasion as a method of earning confidence, and speculation as a method of detecting fabrication.

Therefore, we started a project to create an AI Werewolf (AI Wolf), which plays the Werewolf game in place of a human. In addition, there are several trials for improving AI with game competitions such as Lemonade Stand Game competition [5] and Annual Computer Poker competition [6].

This is a comprehensive project, which aims at the development of not only a game-playing algorithm but also virtual agents and real robots. Many tasks must be solved to achieve the stated objective. Similar to the RoboCup [7] approach in robotics, to solve these tasks, we employ a collective-intelligence approach, which uses competition to improve each player's algorithm. A common platform is indispensable when implementing a collective-intelligence approach. In this paper, we describe the outline of the Werewolf platform for AI (AI Wolf Platform) that we developed as an open-source project. We plan to organize a tournament of AI Wolf in which researchers from various backgrounds can participate freely, with the aim of realizing collective intelligence with the participating researchers.

Section 2 consists of the related studies about incomplete information games and Sect. 3 provides the overview of the werewolf games. Next, the AI Wolf project is defined in Sect. 4, and we describe our analysis for the first competition in Sect. 5. Finally, Sect. 6 provides the conclusion and future proposals.

2   Challenges for Incomplete-Information Games

The development of a game-playing agent has been a challenge from the beginning of AI research [8]. Several two-player board games with perfect information, such as Checkers, Othello, Chess, and Go, have been used for trials by applying several new algorithms [9, 10]. In these games, all information is observable by both players. An AI system must only handle the condition of the board and does not need to determine a competitor's thought processes.

Further, there are several unsolved games in incomplete information game fields. Card games have information that cannot be observed by other players [11]. This is also an important field in AI research. Poker is one of the best-known examples, on which several theoretical analyses have been conducted [12]. Other games, including Bridge and the two-player version of Dou Zi Zhu (a popular game in China), have also been studied [13, 14]. Compared to these games, the Werewolf game requires intelligence to estimate the roles and internal states of other players. Although their information cannot be observed by other players, each player's role in the aforementioned games is determined before the game starts and is known to all players. In contrast, a player's role in the Werewolf game is hidden from the other players and is only revealed at the end of the game. This type of situation requires more intelligence because each player (especially a villager) needs to hold multiple world models for the other players' actions. It also suggests that a stable strategy does not exist because if some action suggests that a player supports the villagers, a werewolf will mimic this action. Inaba analyzed the change in the theory in the online werewolf game called "werewolf bulletin board system (BBS)" [15] for 10 years. In addition, this game requires persuasion of other players. This type of intelligence requires two levels of the Theory of Mind: the expectation of other players' expectations [16]. All these considerations suggest that research on the Werewolf game will lead to several new findings in the field of AI.

2.1 Studies on the Werewolf Game

There have been some studies on the Werewolf (or Mafia) game, including a mathematical analysis [17, 18]. In addition, some researchers have attempted to detect a player's role by using the length of utterances and the number of interruptions of a speaker [19], nonverbal information [20], hand and head motions [21], and the words used in discussions [22].

Some researchers have used the Werewolf game as a study for human–agent interactions. Aylett et al. [23] applied the Werewolf game for educating children on cultural sensitivity. Katagami et al. [20] investigated the effect of nonverbal information in the Werewolf game. However, studies have not attempted to develop playing agents.

To realize Werewolf playing agents, many tasks must be solved. These include the asymmetric diversity of player information, persuasion as a means of earning confidence, and speculation to detect fabrication. These tasks are not generally considered in the field of AI agents.

3 AI Wolf Project

3.1 What Is "Are You a Werewolf?"

Overview

Werewolf is a popular party game played worldwide. Werewolf card sets include "Are You a Werewolf?" and "Lupus in Tabula." The game is still played around the world. Additionally, "Mafia" has an identical game structure but with a much less magic-based theme. "Are you a Werewolf?" is a party game that models a conflict between an informed minority and an uninformed majority. Initially, each player is secretly assigned a role affiliated with one of these teams. There are two phases: night and day. At night, the werewolves "attack" the townsfolk. During the day, surviving players discuss the elimination of a werewolf by voting. The objective of the werewolves is to kill off all the villagers without being killed themselves. The objective of the townsfolk is to ascertain who the werewolves are and to kill them.

There are two techniques for playing Werewolf. The first includes face-to-face play by using the game cards described earlier. The other is to play online using web applications, or a BBS-type platform. For example, large BBS services exist in Japan for playing Werewolf. In fact, there are more than a thousand logs of Werewolf games. Many players still play Werewolf on a BBS. Moreover, some academic studies make use of the BBS game logs. Developing physical robots that can play Werewolf face-to-face is one objective of our project; however, many problems must be solved. Consequently, in this paper, we use a simplified representation of the essence of the game based on a BBS-type Werewolf.

Game Procedures

The roles of all players are allocated randomly. Players are divided into two teams, townsfolk and werewolf teams, according to their roles and the method of winning of their teams. The victory condition for the townsfolk is to kill all the werewolves. For the werewolves, the victory condition is to kill humans such that they become equal or fewer in number to the werewolves. A player fundamentally cannot know the other players' roles because the allocated roles are unpublicized. A basic course of action for the townsfolk players is to discover werewolves through conversation because they do not know who the werewolves are. In contrast, the werewolf players know who the werewolves are. Therefore, a basic course of action for the werewolf players is to engage in various cooperative maneuvering, without the townsfolk knowing about their roles.

The game proceeds in alternating phases of day and night. During the day, all players discuss who the werewolves are. Simultaneously, players who have special abilities (which we discuss later) lead discussions that produce advantages for their respective teams by using the information derived from their abilities. After a certain period, players execute one player who is suspected of being a werewolf, as chosen by majority voting. The executed player then leaves the game and cannot play. During the night, werewolf players can attack a townsfolk team player. The attacked player is killed and is eliminated from the game. In addition, players who have special abilities can use those abilities during the night phase. The day and night phases alternate until the winning conditions are met.

Townsfolk players must be able to detect a werewolf player's lie. In addition, persuading other players by using the information obtained through their special abilities is important. Furthermore, a crucially important point for the werewolf team is to manipulate the discussion to the team's advantage. Occasionally, they must impersonate a role and obfuscate the conditions and evidence.

Roles of Players

There are many variations of the rules and roles of the Werewolf game. Therefore, we use the following basic set of roles for simplification.

- Villager: Townsfolk team. A character in this role has no special ability.
- Werewolf: Werewolf team. Werewolves can attack one townsfolk player during each night phase. They all decide on a single player to attack together with vote, and zero or one villager dies each night. BBS-type game also allows werewolves to talk with each other simultaneously during the day, and we used the same rules in this AI game.
- Seer: Townsfolk team. A seer can inspect a player in every night phase to ascertain whether or not a player is a werewolf.
- Bodyguard: Townsfolk team. A bodyguard can choose a player in every night phase and protect the player against an attack by a werewolf.
- Medium: Townsfolk team. A medium can ascertain whether a player who was executed during the previous day phase was a werewolf.
- Possessed: Werewolf team. Werewolves do not know who is a possessed player. The possessed have no special ability. This role secretly cooperates with werewolves because a werewolf-team victory is also regarded as a victory for possessed players.

3.2 Roadmap of AI Wolf Project

We plan to create AI agents that can play the Werewolf game [9]. It is an incomplete-information game. In addition, the Werewolf game is conducted solely through discussion, and players must use their cognitive faculties completely to win. The symbolization of the Werewolf game is difficult compared to other incomplete-information games such as Poker. This feature requires a different approach than other incomplete-game challenges.

An AI agent requires multiple research areas, such as analyzing the human playing Werewolf, natural language processing, agent technology, and human-agent interaction. Our project consists of not only a sole project team but also of multiple research teams. Figure 1 explains the milestones of the project, the keystone of which is a Werewolf intelligence competition (WIC) that gathers the collective intelligence of people in a program.

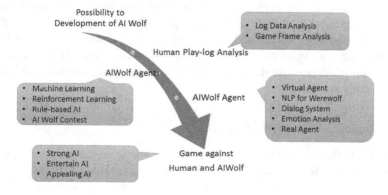

Fig. 1. Project plan of WIC

4 AI Wolf Platform

4.1 Architecture of the AI Wolf Platform

We have been developing the AI Wolf Platform, which is intended to function as an apparatus for evolving AI Wolf agent Game Player algorithms through collective intelligence. The platform consists of the game server and game-player agents (as shown schematically in Fig. 2). These agents connect to the server and play the Werewolf game. Therefore, this platform is built on the client–server architecture. The game server performs the role of game moderator. Moreover, the server controls the network communication between the agents and itself and maintains a log of the games. Game-player agents communicate with the game server via TCP/IP or an internal function-call API. By using the TCP/IP connection, developers can play against other wired player agents. In addition, by using the internal function call, developers can conduct high-speed simulations. The AI Wolf Platform has a communication protocol API between the server and clients. This API is an abstraction layer for the game server and player agents. It facilitates parsing by restricting communication to a specific content format.

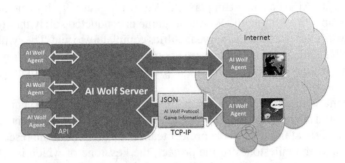

Fig. 2. AI Wolf platform architecture

The game server library is offered by Java. Agent-building libraries are offered by Java, .NET Framework, and Python. Agent programmers need not be concerned about communication between server and client because the communication protocol is wrapped by a library. Agent programmers simply implement the Player interface. All agent classes work by an event-driven method. The server asks clients for their behavior, and clients reply accordingly.

4.2 Development of an AI Wolf Agent

Each agent acts through event-driven systems. The game server sends a request and agents return a response as an action in the Werewolf game. Table 1 shows the requests that are sent from the AI Wolf Server to the agents. Therefore, a game-agent developer must consider only how agents should act when each request arrives. In summary, the developer must implement some method that corresponds to each request.

Table 1. Requests from the AI Wolf Server

Request	Agent action	Reply
Initialize	Initialize for game start	–
DailyInitia lize	Initialize for day Start	–
Finish	Finish the game	–
Name	Return name of the agent	Name
Role	Return the role of an agent	Role
Talk	Talk to other agents	Talk
Whisper	Talk to other werewolves	Whisper
Vote	Choose an agent to be voted	Agent
Divine	Choose an agent to be divined	Agent
Guard	Choose an agent to be guarded	Agent
Attack	Choose an agent to be attacked	Agent

In the Werewolf game, an agent should change its behavior pattern depending on its role. The possible requests differ for each role.

To simplify the accommodation of different roles, the AI Wolf agent library contains the class of AbstractRolePlayer (as shown at the left in Fig. 3). When the developers implement a role, they program the AbstractPlayer class (e.g., in the case of Seer, it would be AbstractSeerPlayer) and assign its class as the function MyRoleAssignPlayer. For example, if we wish to implement a Seer player, we program AbstractSeerPlayer and assign it to MyRoleAssignPlayer, whereupon the agent can act as a seer player. This AbstractRoleAssignPlayer has default behaviors for all roles. The developer should not have to create all the role-behavior algorithms but can instead use default algorithms (or those of other developers) from an early stage.

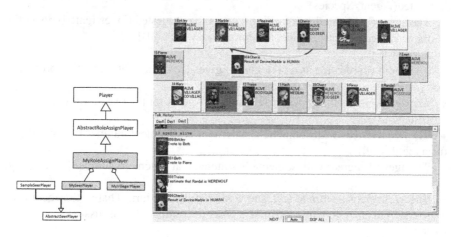

Fig. 3. Class diagram of AbstractRolePlayer (left) and GUI log viewer (right)

Furthermore, we implemented a GUI log viewer (shown at the right in Fig. 3) to help with program debugging. It can be used not only for showing the behavior of agents but also for interactive debugging when programming an agent's behavior.

4.3   AI Wolf Protocol

During the game, agents communicate with other agents using the AI Wolf Protocol, which is a shortened communication protocol designed for AI Wolf. This communication protocol is determined by referring to frequent utterances used in Werewolf BBS. The Werewolf BBS allows a limited number of communications (20 per day for the villagers, and 30 per day for werewolves); this limitation causes shortened symbols. For example, the expressing of a role is called "coming out"' (divulgence), which is shortened to "CO." As such, CO and other designators are used distinctively; we applied this difference to our protocol.

The current version of AI Wolf Platform employs a simple protocol as the first step of the project. This simple protocol permits only limited utterances, such as "I declare as seer" and "I suspect that he is a werewolf." We evaluated the Werewolf BBS logs, in which 50% of the utterances are represented through 10 protocols. Hence, each agent can use the following 10 communication protocols as explained:

- estimate(Agent, Role)
  - An agent expresses its suspicion that [Agent] is [Role].
- comingout(Agent, Role)
  - The agent asserts that [Agent] is [Role].
- divined(Agent, Species)
  - The agent (implicated as a seer) gives the divined result that [Agent] is [Species (human or werewolf)]
- inquested(Agent, Species)
  - The agent (implicated as a medium) gives the inquested (investigated) result that the executed [Agent] is [Species (human or werewolf)]
- guarded(Agent)
  - The agent (implicated as a bodyguard) gives the result that [Agent] is protected.
- vote(Agent)
  - The agent claims that a player will select [Agent] for the execution vote
- agree(day, id)
  - The agent agrees with someone's statement at statement number [id] on [day].
- disagree(day, id)
  - The agent disagrees with someone's statement at statement number [id] on [day].
- skip()
  - The agent skips its turn to talk, and waits for the next turn. That is, the agent waits to listen to an opponent's talk and wishes to continue the discussion.
- over()
  - The agent skips its turn to talk, waits for the next turn, and agrees to finish its discussion the same day.

To ease the development of AI Wolf agents, the platform provides an utterance factory and parser for the protocol.

5   The First WIC

We organized the first WIC at the Computer Entertainment Developers Conference (CEDEC) on August 27, 2015. CEDEC is one of the biggest domestic conferences in Japan for video-game competitions, and is being organized since the last 17 years. More than 30,000 people participate in the conference. Representatives from academic institutions and video-game companies attend and exchange their findings. Moreover, several international research sessions are organized. Thus, we considered that this conference would be a good forum to evaluate our approach.

5.1   Rules of the Competition

We organized preliminary and final competitions. Both competitions were staged according to a BBS-type Werewolf game. Fifteen agents joined one game set, and roles listed in Table 2 were assigned to each agent. One set comprised 100 games, and agents was the same in each set but with different assigned roles. Each agent in the winning team received one point in each game.

Table 2. Roles and agents

Role	Count	Side
Villager	8	Villager
Seer	1	Villager
Medium	1	Villager
Bodyguard	1	Villager
Werewolf	3	Werewolf
Possessed	1	Werewolf

In the final competition, 1,124,890 games were played and 15 agents were assigned their roles randomly in the games. All the games were ranked.

5.2   Participants

Each participating team could submit one agent program. 78 teams joined the competition, and 45 teams submitted programs. Seven agents were rejected because of errors; hence, 38 agents joined the preliminary competition.

Table 3 shows the fraction of students in the competition. More than half of the participants comprised students from universities and other educational organizations. This result suggests that these participants not only joined in with the programming competition, but also focused on the research associated with the Werewolf game (several students also published research on the Werewolf game after the competition).

However, the lower rate of students in the final competition suggests that the professionals, including video programmers, were more proficient than the students.

Table 3. Participating teams

	Total	Students	Fraction of Students
Registered	78	42	0.53
Preliminary	38	24	0.63
Final	15	7	0.47

Figure 4 shows some of the participants during the final competition at CEDEC 2015. More than 200 participants took part in the final competition, and our results were reported through at least five media outlets. In the CEDEC 2015 session, we selected one example from the final competition, and participants explained their algorithms to each other.

Fig. 4. Final competition in CEDEC 2015

5.3   Results of the Competition

The left of Fig. 5 shows the success rates of all 38 teams who took part in the preliminary competition. According to this result, although high-ranking agents generally have higher success rates, most agents have rates that are approximately the same. This may be because Werewolf is a multiple-player game. As such, each agent's contribution toward a win is lower than it would be in a single-player game. The rate of the top-ranked agent is 0.4915, whereas the rate of the bottom-ranked agent is 0.3629. This represents a 13% difference between the strengths of agents. There is no significant difference between the agents ranked 15th and 16th, and participants in this rank border are assigned by luck. We need to improve our rules for the next competition to reflect significant differences at the borders.

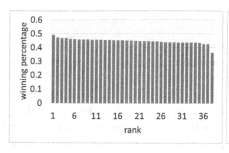

Fig. 5. Success rate of agents in preliminary competition (left) and in final competition (right)

The right of Fig. 5 shows the success rates of all 15 agents who took part in the final competition. We statistically analyzed the difference between two odd ratios. The result suggests that the five top-ranked agents are significantly stronger than the other 10 agents.

5.4 Analysis of the Final Competition

Table 4 shows the types of roles in which an agent was strongest. The results suggest that the top-ranked agents are strong in nearly every role.

Table 4. Ranking for each role

Final rank	Villager rank	Seer rank	Medium rank	Bodyguard rank	Werewolf rank	Possessed rank
1	1	1	7	1	1	1
2	4	2	8	2	3	4
3	3	3	3	8	2	14
4	2	11	12	4	4	5
5	10	7	1	5	5	3
6	9	4	9	10	6	8
7	11	9	4	6	12	2
8	6	10	15	3	10	6
9	14	5	10	7	7	12
10	5	14	2	15	8	11
11	13	6	11	11	13	7
12	8	13	5	12	11	10
13	12	8	13	13	9	15
14	7	15	6	9	14	9
15	15	12	14	14	15	13

First, we calculated the total success rate for each role and the difference between square errors of these rates. Next, we plotted the results using multidimensional scaling. We calculated the distance D_{kl} between points k and l according to the following equation:

$$D_{kl} = \Sigma_i(x_{ik} - x_{il})^2$$

The value of x_{ik} represents the success rate of role k for agent i. Figure 6 (left/right) shows the data from the preliminary/final competition. The success rates for Villager, Medium, and Bodyguard are relatively close, whereas those for Seer, Werewolf, and Possessed are distinctly different. The Seer, Werewolf, and Possessed roles require more specific skills, thus explaining the large distances in the plot in Fig. 6. In the final competition, the Medium and Possessed roles showed different behaviors than in the previous competition. We speculate that players in the final competition wrote more intelligent code than those in the preliminary competition. For a Medium, the highest and lowest scoring agents showed very slight difference, indicating that a Medium does not contribute much toward winning: a result similar to those obtained through statistical analyses of online Werewolf games in Japan. Moreover, some high-ranking players score lower than the low-ranking players in the Possessed role. This suggests that the Possessed role requires certain unique features than the other roles. We speculate that this apparent fact helped players with their programming-resource management from one competition to the next.

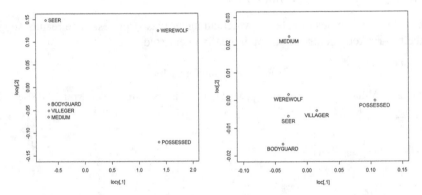

Fig. 6. Plots for success-rate distance of each role (left: preliminary competition; right: final competition).

Lastly, we evaluated the difference between the success rates of agents who did and did not reach the final according to each role (Fig. 7). In all roles, finalists are stronger by a significant difference. In particular, the differences range between 4% and 5% for both Seer and Werewolf roles. This fact may be helpful in suggesting where programmers could best focus AI in the Werewolf game.

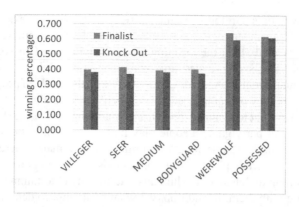

Fig. 7. Success rates of preliminary and finalist agents in each role

5.5   Discussion

There are several trends that are observed in the participating agents.

1. Strong agents tend to be strong in any role.
2. Seer, Werewolf, or Possessed scores might differ from their real ability in the preliminary competition.
3. Although higher-ranked agents tend to score higher, the success rates of Medium and Possessed roles differ from the other roles.
4. All agents in the final competition are significantly stronger than unsuccessful agents in any role. The difference is especially clear for both Seer and Werewolf.

Our findings suggest that these competition trials facilitate a collective-intelligence approach, the findings of which contribute to the analysis of Werewolf games. However, there are still some agents with anomalous behavior, even in the final competition. These difficulties may improve in future challenges. For further evaluation, we want to reflect evaluation methods for other competitions [24, 25].

The first competition was reported by several media outlets (NIKKEI, Game Watch, GPara.com, ASCII/Digital). We believe that our challenge made a good impact on society, and the notion of "lying AIs" stimulated discussions on the role of AI in modern society.

6   Conclusion and Future Work

This paper summarized the AI Wolf project, its competition server, and results from the first WIC. Thirty-eight agents participated in the competition, with 15 agents participating in the finals. The top agent UDON was significantly stronger than all the other agents. The results of the study support the assertion that a competition facilitates in achieving collective intelligence.

The analysis of the relationship between roles and success rates reveals several role-dependent features. Theses agent sources are available on an open-source basis on our site, and agents who outsmart these programs are likely to win in the next competition. This step "evolves" agent strategy, and we expect that this ecosystem will produce new findings for the AI field about the meaning of communication. Our source codes are completely open. The server codes are available on GitHub (https://github.com/aiwolf/), and the final agent source codes are available on the AI Wolf project site (http://aiwolf.org/). We are now planning a second competition for CEDEC 2016.

In the future, we want to evaluate the value of entertainment of the Werewolf game. Werewolf is a type of party game. People play party games for not only winning and losing but also enjoying the game and its communication element. Therefore, an AI for Werewolf that can "entertain players" must be established. We assume that if the agents are sufficiently strong to defeat all the humans without any entertainment, most people will avoid playing with them. In addition, we want to attempt to determine the source of pleasure in competitive games. Sometimes, the value of entertainment can be understood from a game's rules. However, in many cases the action of opponents becomes a factor in determining the pleasure derived from a game. Therefore, to understand "why games entertain players," it is necessary to consider the interaction between the opponent's behavior and the game's rules. In certain situations, players may enjoy a game. Such situations include competitive games between human players. In such a case, the purpose of AI can be assumed to be the generation of those pleasant situations. Therefore, the AI must act to change the game environment.

Acknowledgements. This work was partially supported by Hayao Nakayama Foundation for Science & Technology and Culture, Foundation for Fusion of Science and Technology, and JSPS KAKENHI Grant Number 26118006. We also want to say thanks for Computer Entertainment Developers Conference and Japan Society for Artificial Intelligence.

References

1. Silver, D., Huang, A., Maddison, C.J., Guez, A., Sifre, L., van den Driessche, G., Schrittwieser, J., Antonoglou, I., Panneershelvam, V., Lanctot, M., Dieleman, S., Grewe, D., Nham, J., Kalchbrenner, N., Sutskever, I., Lillicrap, T., Leach, M., Kavukcuoglu, K., Graepel, T., Hassabis, D.: Mastering the game of Go with deep neural networks and tree search. Nature **529**, 484–489 (2016)
2. Genesereth, M., Love, N., Pell, B.: General game playing: overview of the AAAI competition. AI Mag. **26**, 1–16 (2005)
3. Bowling, M., Burch, N., Johanson, M., Tammelin, O.: Heads-up limit hold'em poker is solved. Science **347**, 145–149 (2015)
4. Karakovskiy, S., Togelius, J.: The Mario AI benchmark and competitions. IEEE Trans. Comput. Intell. AI Games **4**, 55–67 (2012)
5. Zinkevich, M.A., Bowling, M., Wunder, M.: The lemonade stand game competition. ACM SIGecom Exchanges **10**, 35–38 (2011)
6. ACPC: Computer Poker Competition. www.computerpokercompetition.org/
7. Kitano, H., Asada, M., Kuniyoshi, Y., Noda, I., Osawa, E.: RoboCup: the robot world cup initiative. In: Proceedings of the First International Conference on Autonomous Agents - AGENTS 1997, pp. 340–347. ACM Press, New York (1997)

8. Abramson, B.: Control strategies for two-player games. ACM Comput. Surv. **21**, 137–161 (1989)
9. Krawiec, K., Szubert, M.G.: Learning N-tuple networks for othello by coevolutionary gradient search. In: Proceedings of the 13th Annual Conference on Genetic and Evolutionary Computation - GECCO 2011, pp. 355–362. ACM Press, New York (2011)
10. Gelly, S., Kocsis, L., Schoenauer, M., Sebag, M., Silver, D., Szepesvári, C., Teytaud, O.: The grand challenge of computer Go. Commun. ACM **55**, 106–113 (2012)
11. Ganzfried, S., Sandholm, T.: Game theory-based opponent modeling in large imperfect-information games. In: International Conference on Autonomous Agents and Multiagent Systems, pp. 533–540. International Foundation for Autonomous Agents and Multiagent Systems (2011)
12. Billings, D., Burch, N., Davidson, A., Holte, R., Schaeffer, J., Schauenberg, T., Szafron, D.: Approximating game-theoretic optimal strategies for full-scale poker. In: International Joint Conference on Artificial Intelligence, pp. 661–668 (2003)
13. Ginsberg, M.L.: GIB: imperfect information in a computationally challenging game. J. Artif. Intell. Res. **14**, 303–358 (2001)
14. Whitehouse, D., Powley, E.J., Cowling, P.I.: Determinization and information set Monte Carlo tree search for the card game Dou Di Zhu. In: 2011 IEEE Conference on Computational Intelligence and Games (CIG 2011), pp. 87–94. IEEE (2011)
15. Inaba, M.I., Toriumi, F.U., Takahashi, K.E.: The statistical analysis of werewolf game data. In: Proceedings of Game Programming Workshop 2012, pp. 144–147 (2012) (in Japanese)
16. Dias, J., Reis, H., Paiva, A.: Lie to me: virtual agents that lie. In: Proceedings of the 2013 International Conference on Autonomous Agents and Multi-agent Systems, pp. 1211–1212 (2013)
17. Braverman, M., Etesami, O., Mossel, E.: Mafia: a theoretical study of players and coalitions in a partial information environment. Ann. Appl. Probab. **18**, 825–846 (2008)
18. Yao, E.: A theoretical study of Mafia games (2008). Arxiv Prepr. arXiv:0804.0071
19. Chittaranjan, G., Hung, H.: Are you a werewolf? Detecting deceptive roles and outcomes in a conversational role-playing game. In: 2010 IEEE International Conference on Acoustics, Speech and Signal Processing, pp. 5334–5337. IEEE (2010)
20. Katagami, D., Kanazawa, M., Toriumi, F., Osawa, H., Inaba, M., Shinoda, K.: Movement design of a life-like agent for the werewolf game. In: IEEE International Conference on Fuzzy Systems, pp. 982–987 (2015)
21. Xia, F., Wang, H., Huang, J.: Deception detection via blob motion pattern analysis. In: Paiva, A.C.R., Prada, R., Picard, R.W. (eds.) ACII 2007 Proceedings of the 2nd International Conference on Affective Computing and Intelligent Interaction, pp. 727–728. Springer, Heidelberg (2007)
22. Zhou, L., Sung, Y.: Cues to deception in online chinese groups. In: Proceedings of the 41st Annual Hawaii International Conference on System Sciences (HICSS 2008), p. 146. IEEE (2008)
23. Aylett, R., Hall, L., Tazzyman, S., Endrass, B., André, E., Ritter, C., Nazir, A., Paiva, A., Höfstede, G., Kappas, A.: Werewolves, cheats, and cultural sensitivity. In: Autonomous Agents and Multi-Agent Systems, pp. 1085–1092. International Foundation for Autonomous Agents and Multiagent Systems (2014)
24. White, M., Bowling, M.: Learning a value analysis tool for agent evaluation. In: IJCAI 2009 Proceedings of the 21st International Joint Conference on Artificial Intelligence, pp. 1976–1981 (2009)
25. Davidson, J., Archibald, C., Bowling, M.: Baseline: practical control variates for agent evaluation in zero-sum domains. In: Proceedings of the 2013 International Conference on Autonomous Agents and Multi-agent Systems, pp. 1005–1012 (2013)

Semantic Classification of Utterances in a Language-Driven Game

Kellen Gillespie[1,2], Michael W. Floyd[2(✉)], Matthew Molineaux[2],
Swaroop S. Vattam[3], and David W. Aha[4]

[1] Amazon.Com, Inc., Seattle, WA, USA
kelleng@amazon.com
[2] Knexus Research Corporation, Springfield, VA, USA
michael.floyd@knexusresearch.com
matthew.molineaux@knexusresearch.com
[3] MIT Lincoln Laboratory (Group 52), Lexington, MA, USA
swaroop.vattam@ll.mit.edu
[4] Naval Research Laboratory (Code 5514), Washington, DC, USA
david.aha@nrl.navy.mil

Abstract. Artificial agents that interact with humans may find that understanding those humans' plans and goals can improve their interactions. Ideally, humans would explicitly provide information about their plans, goals, and motivations to the agent. However, if the human is unable or unwilling to provide this information then the agent will need to infer it from observed behavior. We describe a goal reasoning agent architecture that allows an agent to classify natural language utterances, hypothesize about human's actions, and recognize their plans and goals. In this paper we focus on one module of our architecture, the *Natural Language Classifier*, and demonstrate its use in a multiplayer tabletop social deception game, *One Night Ultimate Werewolf*. Our evaluation indicates that our system can obtain reasonable performance even when the utterances are unstructured, deceptive, or ambiguous.

Keywords: Semantic classification · Social deception game · Tabletop game · Goal reasoning

1 Introduction

Agents that interact with humans, cooperatively or competitively, can benefit from understanding those humans' plans and goals. By having this information, the agent can more effectively assist a human teammate or thwart an adversarial human. While in some circumstances a human may directly and concisely provide its plans and goals, it is often more realistic that the agent will need to infer this information based on the human's behavior. In this work, we consider a particular problem domain where humans do not unambiguously share this type of information, and will often attempt to intentionally conceal it through deception.

© Springer International Publishing AG 2017
T. Cazenave et al. (Eds.): CGW 2016/GIGA 2016, CCIS 705, pp. 116–129, 2017.
DOI: 10.1007/978-3-319-57969-6_9

In this paper, we describe our architecture for an agent that classifies natural language utterances to hypothesize about humans' plans and goals. We have previously shown that such an agent can successfully predict squad members' goals in a military domain (Gillespie et al. 2015). However, deploying the agent in a social deception game adds the following complexities:

- **Human cooperation:**
  - *Military domain:* The humans are squad members working in collaboration with the agent.
  - *Social deception game:* The humans can be teammates of the agent but can also be neutral or adversaries.
- **Language:**
  - *Military domain:* The fixed-vocabulary language is highly constrained.
  - *Social deception game:* There are minimal constraints on the language.
- **Clarity of utterances:**
  - *Military domain:* The utterances will be direct, concise, and unambiguous.
  - *Social deception game:* The utterances may be incomplete, ambiguous, incorrect, or deceptive. Additionally, some utterances may have no relevance to the game (e.g., casual conversation among players).

Although our focus has been on military scenarios and social deception games, the ability to reason about goals from natural language is also relevant in other domains such as those involving negotiations, diplomacy, and legal reasoning.

While we describe the entire agent architecture in Sect. 2, our focus in this paper is on the module that allows the agent to classify the semantic meaning of each utterance. Section 3 provides an introduction to the social deception game we use, One Night Ultimate Werewolf, and Sect. 4 presents our approach for extracting information from in-game utterances. In Sect. 5, we describe an evaluation using logs of actual gameplay and show that the agent is able to classify several key aspects of each utterance. We examine related work in Sect. 6 and present future research directions in Sect. 7.

2 Agent Architecture

Our agent interprets and responds to its environment via a five-step goal reasoning process (Klenk et al. 2013; Aha 2015). This process allows an agent to dynamically refine its goals in response to unexpected external events or opportunities, and enact plans to accomplish those goals. The agent's decision cycle is shown in Fig. 1 and has five primary components:

1. **Natural Language Classifier:** This module listens for natural language *utterances* (i.e., spoken language) in the environment and attempts to extract semantic meaning from the utterances. For each utterance received, the module outputs a multi-label *classification of the utterance*.
2. **Explanation Generator:** This module uses the *classified utterances* and *environmental observations* (i.e., the current state of the environment) to generate possible explanations for what has occurred in the environment (Molineaux and Aha 2015).

The explanation contains, in part, the agent's hypothesis as to what actions each other entity (e.g., humans, robots, or other agents) in the environment must have performed for the environment to have changed from its prior state to the current state. As more classified utterances and state observations are received, the Explanation Generator further refines its explanation. The most likely *actions* for each entity are output.

3. **Plan Recognizer:** For each entity in the environment, the Plan Recognizer receives a sequence of *actions* that the entity may have performed (i.e., one action in the sequence every time the Explanation Generator produces output). The Plan Recognizer uses the sequence of actions to identify the entity's plan (Vattam et al. 2014). The Plan Recognizer assumes that each plan achieves a goal, so the recognized plan can be used to identify the entity's current goal. This module outputs the recognized *goal* of each entity in the environment.

4. **Goal Selector:** This module monitors for any changes in the *goals of the entities* or external events, and can modify the agent's goal in response. This allows the agent to dynamically respond to any unexpected behaviors or opportunities (i.e., the agent changes its goal to better respond to other entities' goals). The output of this module is the *agent's goal* (even if the goal is unchanged).

5. **Plan Generator:** If the *agent's goal* has changed, the Plan Generator generates a new plan for the agent to perform. The plan generator also monitors the progress of the current plan to determine if it is necessary to repair the plan or generate a new plan. The output of this module are the *actions* (of the plan) that the agent is attempting to perform.

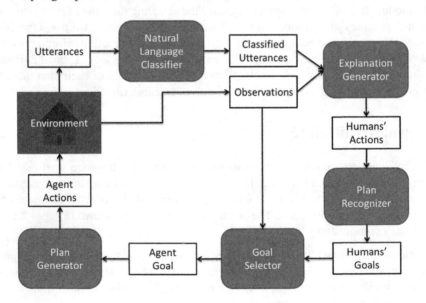

Fig. 1. Decision cycle of the agent

In this paper we focus exclusively on the Natural Language Classifier and how it generates classified utterances from unconstrained natural language.

3 Background: One Night Ultimate Werewolf

The domain we are examining is a tabletop social deception game called *One Night Ultimate Werewolf*[1] (Bezier Games 2016). We chose Ultimate Werewolf because players interact using unconstrained natural language, have a variety of goals, work under hidden information, and actively engage in deception.

In the game, players are randomly assigned *roles* that place them into three competing factions: *Villagers, Werewolves,* and the *Tanner*. The goal of the Villagers is to identify which players are Werewolves, the goal of the Werewolves is to avoid detection, and the goal of the Tanner is to convince the Villagers that it is a Werewolf. We constrained the game to five players and eight possible roles (i.e., five roles will be assigned and three will be unused), with some roles granting special abilities. The roles we use are: *Werewolf* (x2), *Mason* (x2), *Generic Villager* (x2), *Seer,* and *Tanner*. The Werewolf roles are part of the *Werewolves faction,* the Tanner is part of the *Tanner faction,* and all remaining roles are part of the *Villagers faction*. The three unused role cards are placed, face down, on the table.

The game proceeds as follows:

1. **Role assignment:** Each player receives a *role card* with an assigned role printed on it. After viewing their role, the player then places the card face down in front of them. They may not view their card again[2].
2. **Special abilities:** An external moderator oversees this portion of the game:
   (a) The moderator instructs all players to close their eyes.
   (b) The moderator instructs all Werewolves to open their eyes, identify the other Werewolves (if any), and close their eyes. If only one Werewolf opens their eyes, they may look at one of the unused role cards.
   (c) The moderator instructs all Masons to open their eyes, identify the other Masons (if any), and close their eyes.
   (d) The moderator instructs the Seer to open their eyes. The Seer may look at the role card of one other player or two of the unused role cards. The Seer then closes their eyes.
   (e) The moderator instructs all players to open their eyes again.
3. **Information gathering:** The players have several minutes to attempt to gather information about the other players. There is no turn-taking so players can speak as much or as little as they wish. Similarly, there are no constraints on what is discussed or the vocabulary used.

---

[1] We will refer to the game as *Ultimate Werewolf* for the remainder of the paper.
[2] Although viewing your role again does not influence our game, in some versions of Ultimate Werewolf a player's role can be switched without their knowledge.

4. **Shooting phase:** Each player chooses one other player to "shoot" and players announce their choices simultaneously. The player who is shot by the most other players "dies". In the event of a tie, all players tied for the most shots die.

5. **Declaring winners:**
   (a) If the Tanner dies, the Tanner wins (regardless of which other players die). Otherwise, the Tanner loses.
   (b) If at least one Werewolf dies, the Villagers faction wins (regardless of the Tanner's fate). Otherwise, they lose.
   (c) If the Tanner does not die and no Werewolves die, the Werewolves faction wins. Otherwise, the Werewolves lose.

Each player knows their own role and, depending on their special ability, may have more information as well (i.e., from special abilities). The Werewolves and Masons know information about other members of their faction; the Seer may know the role of any one other player; and a lone Werewolf or the Seer may know either 1 or 2 unused roles. Players with the Generic Villager role have no special abilities, so they have less information than other players.

4   Multilabel and Multiclass Semantic Classification

The Natural Language Classifier receives as input each natural language *utterance* that it can sense in the environment. Each utterance represents a continuous unit of speech with a distinct beginning and ending (e.g., *"I think you are a werewolf."* or *"Did you look at anyone's role?"*). Utterances are encoded using a bag-of-words representation. An utterance u is a set containing each word w in the utterance:

$$u = \{w_a, w_b, \ldots\}$$

For example, *"I think you are a werewolf."* would be represented as $\{'I',' think',' you',' are',' a',' werewolf'\}$. We classify each utterance along nine different dimensions using a set of parallel classifiers. The classification tasks and their associated class labels are listed below:

- **Purpose:** The general type of utterance being made.
  - **Classes:** *claim* (make a factual claim), *question* (ask a question), *hypothesis* (pose a hypothesis), *suggest-target* (suggest a target to shoot), *self-explain* (explain the player's behavior to the group), *other* (an utterance that does not fall under any of the other classes).
- **Address-type:** The size of the group the utterance was addressed to.
  - **Classes:** *everyone* (the utterance was directed at all or most of the players), *one-person, two-people.*
- **Addressee:** Whether an utterance is directed to a specific player. This classification task is complementary to Address-type (i.e., a *known* Addressee only occurs when the Address-type is *one-person* or *two-people*).
  - **Classes:** *known* (the utterance directly addresses one of the players), *none* (no specific player is addressed).

- **Subject:** The subject matter discussed in the utterance.
  - **Classes:** *starting-role* (a player's role when they viewed their role card), *unused-role* (roles that were not assigned to anyone), *starting-role-group* (a subgroup of possible roles for a player), *role-observe-performer* (whether a player has a role that allows the observation of other players' roles), *role-observe-target* (whether a player had their role observed by another player), *divulge* (a player provides information about themselves to other players), *statement* (the utterance is in regards to a previously made statement), *shoot-target* (discusses targeting a player for shooting).
- **Target-role:** The role being discussed in the utterance.
  - **Classes:** *none* (no role is being discussed), *unknown* (a role is being discussed but the exact role is not known), *Seer*, *Werewolf*, *Villager*, *Mason*, *Tanner*.
- **Target-role-group:** The subgroup of roles is being directly discussed.
  - **Classes:** *none*, *villagers*, *non-villagers*, *paired-roles* (roles, either Masons or Werewolves, which can view the other members with the same role).
- **Target-player:** The player being discussed in the utterance.
  - **Classes:** *known* (directly referring to one of the players), *unknown* (a player is discussed but the exact player is unknown), *none* (no player is discussed).
- **Target-position:** The presence and location of an unused role card on the table (e.g., a card viewed by the Seer, knowledge of an unused role because there were no other Werewolves).
  - **Classes:** *one-unknown* (a role is unused but its position is unknown), *two-unknown* (two roles are unused but their positions are unknown), *three-unknown* (three roles are unused but their positions are unknown), *left* (the leftmost unused role card), *middle* (the middle unused role card), *right* (the right unused role card), *none* (no unused role is mentioned).
- **Negation:** Whether a statement is positive (e.g., something happened or is true) or negative (e.g., something did not happen or is not true).
  - **Classes:** *positive*, *negative*.

4.1 Classifiers

We examine three methods for training the classifiers used by the Natural Language Classifier: *Frequency*, *Probabilistic*, and *Probabilistic Frequency*. All three methods use a dictionary of known words. If there are N known words, the dictionary $dict$ will contain N entries ($dict = \langle w_1, w_2, \ldots, w_N \rangle$). Each utterance u is filtered to remove stop words and converted to a vector v_u of length N ($v_u = \langle m_1, m_2, \ldots, m_N \rangle$). The ith element in v_u (i.e., m_i) contains the multiplicity in the utterance of the ith element in $dict$ (i.e., w_i). For example, if the 3rd word in the dictionary is '*werewolf*' and the word '*werewolf*' occurred in the utterance once, the 3rd element of v_u would be 1.

The three classification methods learn *classification vectors* from a set of labelled training utterances. Like the utterance vectors, the classification vectors are of length N (i.e., classification vector ($cv = \langle s_1, s_2, \ldots, s_N \rangle$)). For each classification task, the training examples are partitioned by class and one classification vector is learned for each class (e.g., for the *Negation* task the training examples are partitioned into one set with the *positive* label and one set with the *negative* label). The three methods generate classification vectors as follows:

Frequency

All utterance vectors from a partition are summed. If the utterance vectors from class C are in partition p_C, classification vector cv_C^{freq} for that class is:

$$cv_C^{freq} = \sum_{v_{u_i} \in p_C} v_{u_i}$$

Since each utterance vector encodes the number of times each word appeared in the utterance, the classification vector contains the total number of times each word appeared for a given class.

Probabilistic

The Probabilistic classification vector cv_C^{prob} is computed by dividing each element of the Frequency classification vector by the number of utterances in the partition:

$$cv_C^{prob} = \frac{cv_C^{freq}}{|p_C|}$$

This classification vector represents what percentage of utterances in the partition contained each word.

Probabilistic Frequency

The Probabilistic Frequency classification vector cv_C^{pf} is calculated using both the Frequency and Probabilistic classification vectors. A new classification vector is created such that the ith element is the product of the ith elements in the Frequency and Probabilistic classification vectors:

$$cv_C^{pf} = \langle s_{C,1}^{freq} \times s_{C,1}^{prob}, s_{C,2}^{freq} \times s_{C,2}^{prob}, \ldots, s_{C,N}^{freq} \times s_{C,N}^{prob} \rangle$$

4.2 Classification

An input utterance is classified by the Natural Language Classifier using the learned classification vectors. If a classification task l has a set of possible labels C_l, the Natural Language Classifier computes the dot product between the utterance vector and each of the classification vectors for that classification task (e.g., to find the *Negation* classification, only the classification vectors for the *positive* and *negative* classes are used). The associated label of the classification vector that maximizes that value is assigned to the utterance:

$$label_l = \text{argmax}_{C_i \in C_l} v_u \cdot cv_{C_i}$$

In the Ultimate Werewolf domain, nine labels are assigned to each input utterance.

5   Evaluation

In our empirical evaluation we assess whether the agent can correctly classify natural language utterances using multilabel and multiclass semantic classification. Using data from real games of Ultimate Werewolf, our results show that our agent can extract important semantic information from utterances without limiting the language of players.

5.1   Data Collection

We collected data from eight games of Ultimate Werewolf, with each game being played by five human players. The same five players participated in all eight games. In addition to the rules described in Sect. 3, the players were also encouraged to use proper names when referring to each other. This was done because the agent only has access to the audio of the game (i.e., it cannot see who a player is facing when speaking). However, this was not strictly enforced so there are instances where the players use pronouns. No other limitations were placed on vocabulary, utterance structure, conversation ordering, or topics of discussion.

Audio was recorded for each game along with the players' roles, special ability actions (e.g., if they viewed another player's role), and shooting targets. Each recording was manually transcribed and separated into the individual utterances. The mean number of utterances per game was 49.1, with a minimum of 36 and a maximum of 69. Each utterance was manually labelled for each of the nine classification tasks. The labelling was done by a third party (i.e., not the players themselves), so it represents how an external observer would classify each utterance rather than a player's intended meaning (e.g., how the observer interpreted ambiguous statements).

5.2   Experimental Setup

Evaluation was performed using leave-one-out testing (i.e., each run used seven annotated game transcripts for training and one for testing). The utterances from the testing transcript were given as input to the agent. The performance of the agent (i.e., how well its classification matched the annotated classes of the utterance) was measured for each of the nine classification tasks. We used the F_1 score to measure performance $(F_1 = 2\dfrac{precision \times recall}{precision + recall})$. The three classification methods described in Sect. 4 were evaluated: *Frequency*, *Probabilistic*, and *Probabilistic Frequency*. The results from these three classification approaches were also compared to a baseline that randomly classifies each utterance (referred to as *Random* in our results).

5.3 Results

The results for each of the nine classification tasks and the overall performance are shown in Figs. 2 and 3. The Probabilistic and Probabilistic Frequency approaches outperformed the baseline over all classification tasks and outperformed the Frequency approach over all tasks except Target-role-group (i.e., all three approaches achieved similar results for this task). Other than the Target-role task (where Probabilistic Frequency performed better), and Purpose and Target-role-group (where they performed similarly), the Probabilistic method outperformed the Probabilistic Frequency method. The Frequency approach performed poorly, underperforming the Random baseline in six of the classification tasks and recording a lower average F_1-score.

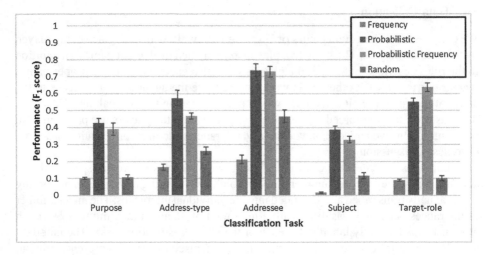

Fig. 2. Classification performance for the *Purpose*, *Address-type*, *Addressee*, *Subject*, and *Target-role* tasks

5.4 Discussion

The classification tasks have between two and eight classes each (with a median of 4). We observed an inverse correlation between the number of classes and agent performance. The two classification tasks that do not follow this inverse correlation are the Target-role and Target-role-group tasks. Target-role has seven classes but the agent performed better than expected on this task. The primary reason for this is because the utterances contain keywords (i.e., the name of the role) that make them easy to classify. In contrast, the agent performed poorly on the Target-role-group task, which has only four classes. This is because the agent has difficulty determining if an utterance is explicitly discussing one of the groups or only implicitly referencing the group by mentioning one of the roles in that group. This is especially prevalent since the players use group names that are similar to role name. For example, "*I think you are one of the villagers*" would be classified as *villagers* (i.e., it discusses the villagers group) whereas

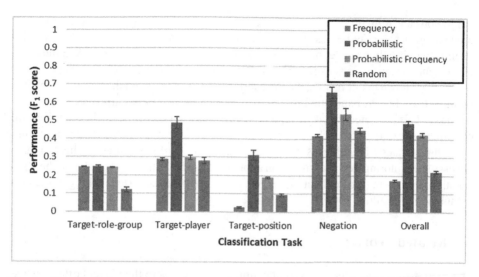

Fig. 3. Classification performance for the *Target-role-group*, *Target-player*, *Target-position*, *Negation*, and *Overall* tasks

"*I think you are the Villager*" would be classified as *none* (i.e., a role is discussed, not an entire group).

The classes are highly imbalanced given the wide range of possible utterances. In our dataset, between 45% and 96% of utterances belong to the majority class ($\mu = 69\%$) and between 0.5% and 28% of the utterances belong to the least frequent class ($\mu = 7\%$). While this imbalance affects all three classification methods, it is the primary reason the Frequency method performs poorly. For each class, the Frequency method counts the number of times each word appears in the training examples. This causes classes with more training examples (i.e., the majority class) to have higher frequency values and therefore be more likely to be the labelled class of an input utterance. Even if a specific word is a strong indication that an utterance should be labelled as the minority class, if that word appears occasionally in the majority class it can cause the classifier to label the utterance as the majority class. The Probabilistic and Probabilistic Frequency approaches help mitigate the class imbalance problem by taking into account the percentage of training examples that contain each word rather than just the number of times a word occurs. However, as with the Frequency approach, they also suffer from having very few training examples for some classes (e.g., some classes only have a single example in the dataset). Additionally, some classes have such a wide range of different utterances (e.g., non-game talk amongst the players) that it makes it difficult to learn a model for that class even if a significant number of examples are available.

Our results, while an improvement over the baseline, fall well short of ideal performance. Given the difficulty of the problem (i.e., unconstrained text, rapid changes in topics, highly unbalanced data, ambiguity), we expected the agent to have difficulty classifying the utterances but are unsure what performance is necessary for the remaining components (i.e., how erroneous the classifications can be before the Explanation Generator and Plan Recognizer fail). Even for a human annotator, the utterances were often highly ambiguous and difficult to classify. While the agent should ideally accurately predict all nine categories, it may be possible that the remaining modules can achieve reasonable results even if only a subset of each utterance's classifications is correct. We intend to investigate the system's sensitivity to classification performance in future work.

As was shown in our results, the Probabilistic method achieved the best performance on most tasks but Probabilistic Frequency performed best on the Target-role classification task. This indicated that it will likely be necessary to determine the best performing classification strategy on each task or use an ensemble approach rather than committing to a single strategy for all tasks. Given our current level of performance, this will also necessitate exploring new classification approaches and taking steps to manage the class imbalance problem (e.g., collect more data, balance the dataset, use label regularization (Mann and McCallum 2007)).

6 Related Work

Our work focuses on utterance classification in a game where the players often engage in deception. Although we do not attempt to identify which utterances or players are deceptive, related work in deception detection often addresses similar problems. Deception detection in conversational games has been approached using textual cues (Zhou and Sung 2008) (e.g., word selection, utterance duration, utterance complexity), vocal cues (Chittaranjan and Hung 2010) (e.g., pitch, pauses, laughter), and visual cues (Raiman et al. 2011) (e.g., head and arm movements). These systems are designed to classify players as truthful or deceptive, and use that information to identify players with deceptive roles (e.g., werewolves). However, while collecting experimental data we observed that even players with roles that should not require deception (e.g., villagers) actively engage in deception and omission. Since nearly all players engage in deception, it becomes more important to identify when they are being deceptive and why they are being deceptive.

Network analysis has been used to identify groups of players with similar patterns of behavior (Yu et al. 2015). The statements made by each player are used to determine their attitudes toward other players (e.g., a positive attitude if they regularly defend another player or a negative attitude if they regularly accuse another player) and players are clustered based on their attitudes. The underlying assumption is that deceptive players will have positive attitudes toward other deceptive players while having negative attitudes toward other players. In our domain, even the most common roles (e.g., Werewolf, Mason, and Generic Villager) only have at most two players with those roles. If a player knows of another player with the same role (i.e., using a special ability), they often avoid displaying a positive attitude toward that player since it can arouse suspicion.

Azaria et al. (2015) have developed an agent that is able to identify deception, convince other players of the deception, and avoid raising suspicions about their own behavior. The agent participates in a simplified social deception game where a single pirate has to deceive three non-pirates in order to steal treasure. The primary differences between their work and our own are that their game uses structured sentences rather than free text, the game is less complex (i.e., fewer roles and player goals), and their system is focused on identifying deception rather than a player's plan or role.

Orkin and Roy (2010) use sequences of utterances and actions to predict a player's behavior in a restaurant simulation game. Due to the number of utterances possible using free-form text, they had relatively poor performance when training with 8-10 game logs compared to 30-100 game logs. This is similar to our own evaluation where many of the classes had few training instances. They found that increasing the number of training logs increased performance but required significant annotation time (approximately 56 h). In the AutoTutor Intelligent Tutoring System (Olney et al. 2003), utterances are used to determine when initiative has changed and determine the needs of the student. For example, certain utterances indicate the student has switched from providing responses to being stuck or asking questions. This can be thought of as a simplified version of plan recognition, where the student has three plans: *respond*, *ask questions*, or *do nothing*. However, only a single utterance is used for each classification, rather than the entire sequence of utterances.

Vázquez et al. (2015) have studied the reaction of human players when a robotic player participates in a social deception game. The robot has the appearance of autonomy but is actually controlled by an unseen human. Although this differs from our own goal of an autonomous player, it does demonstrate that humans are open to playing social deception games with robotic participants.

7 Conclusions and Future Work

We described our architecture for an agent that uses unstructured natural language utterances to reason about the plans and goals of humans. In this paper, we focus on one module of this architecture, the Natural Language Classifier, and examine its ability to classify utterances in a multiplayer tabletop social deception game. Our previous work (Gillespie et al. 2015) described the application of our agent architecture in a military domain. However, in this paper we chose to examine a social deception game because it posed several interesting challenges, including less constrained language, deception, and ambiguity.

The Natural Language Classifier extracts information from each utterance by assigning labels according to nine distinct classification tasks. We studied its ability using three supervised learning methods for these tasks. We evaluated it in the social deception game Ultimate Werewolf using logs of eight games played by human players. We found that classification that considers only word frequency performed poorly, whereas the other two classification methods achieved reasonable results and outperformed our baseline.

Our principal area of future work is to integrate the Natural Language Classifier with the other components of the agent architecture and evaluate the agent's overall performance. We performed such an evaluation in a military domain, but performing this integration for Ultimate Werewolf will require a better understanding of the minimum performance necessary during utterance classification. Currently, we have a limited corpus of training data that was collected from a single set of players. Different players are likely to use different utterances and a different vocabulary, so it will be important to collect data from a variety of players. Additionally, we plan to allow the agent to observe games of Ultimate Werewolf and make predictions about player roles, identify deception, and learn the motivations of individual players.

References

Aha, D.W. (ed.): Goal Reasoning: Papers from the ACS Workshop (Technical Report GT-IRIM-CR-2015-001), Atlanta, USA. Georgia Institute of Technology, Institute for Robotics and Intelligent Machines (2015)

Azaria, A., Richardson, A., Kraus, S.: An agent for deception detection in discussion based environments. In: Proceedings of the Eighteenth ACM Conference on Computer Supported Cooperative Work & Social Computing, Vancouver, Canada, pp. 218–227. ACM (2015)

Bezier Games. One night ultimate werewolf (2016). beziergames.com/collections/all-games/products/one-night-ultimate-werewolf. Accessed

Chittaranjan, G., Hung, H.: Are you a werewolf? detecting deceptive roles and outcomes in a conversational role-playing game. In: Proceedings of the IEEE International Conference on Acoustics, Speech, and Signal Processing, Dallas, USA, pp. 5334–5337. IEEE (2010)

Gillespie, K., Molineaux, M., Floyd, M.W., Vattam, S.S., Aha, D.W.: Goal reasoning for an autonomous squad member. In: Aha, D.W. (ed.) Goal Reasoning: Papers from the ACS Workshop (Technical Report). Atlanta, USA. Georgia Institute of Technology, Institute for Robotics and Intelligent Machines (2015)

Klenk, M., Molineaux, M., Aha, D.W.: Goal-driven autonomy for responding to unexpected events in strategy simulations. Comput. Intell. **29**(2), 187–206 (2013)

Mann, G.S., McCallum, A.: Simple, robust, scalable semi-supervised learning via expectation regularization. In: Proceedings of the Twenty-Fourth International Conference on Machine Learning, Corvallis, USA, pp. 593–600. ACM (2007)

Molineaux, M., Aha, D.W.: Continuous explanation generation in a multi-agent domain. In: Proceedings of the Third Conference on Advances in Cognitive Systems. Cognitive Systems Foundation, Atlanta, USA (2015)

Olney, A.M., Louwerse, M., Matthews, E., Marineau, J., Hite-Mitchell, H., Graesser, A.C.: Utterance classification in AutoTutor. In: Proceedings of the Workshop on Building Educational Applications Using Natural Language Processing at the Human Language Technology Conference of the North American Chapter of the Association for Computational Linguistics, Edmonton, Canada (2003)

Orkin, J., Roy, D.: Semi-automated dialogue act classification for situated social agents in games. In: Proceedings of the Agents for Games & Simulations Workshop at the Ninth International Conference on Autonomous Agents and Multiagent Systems, Toronto, Canada (2010)

Raiman, N., Hung, H., Englebienne, G.: Move, and I will tell you who you are: Detecting deceptive roles in low-quality data. In: Proceedings of the Thirteenth International Conference on Multimodal Interfaces, Alicante, Spain, pp. 201–204. ACM (2011)

Vattam, S.S., Aha, D.W., Floyd, M.: Case-based plan recognition using action sequence graphs. In: Lamontagne, L., Plaza, E. (eds.) ICCBR 2014. LNCS (LNAI), vol. 8765, pp. 495–510. Springer, Cham (2014). doi:10.1007/978-3-319-11209-1_35

Vázquez, M., Carter, E.J., Vaz, J.A., Forlizzi, J., Steinfeld, A., Hudson, S.E.: Social group interactions in a role-playing game. In: Proceedings of the Tenth Annual ACM/IEEE International Conference on Human-Robot Interaction, Portland, USA, pp. 9–10. ACM (2015)

Yu, D., Tyshchuk, Y., Ji, H., Wallace, W.A. Detecting deceptive groups using conversations and network analysis. In: Proceedings of the Fifty-Third Annual Meeting of the Association for Computational Linguistics, Beijing, China, pp. 857–866. ACL (2015)

Zhou, L., Sung, Y.-W.: Cues to deception in online chinese groups. In: Proceedings of the Forty-First Hawaii International Conference on Systems Science, Waikoloa, USA (2008)

General Intelligence in Game-Playing Agents 2016

Optimizing Propositional Networks

Chiara F. Sironi$^{(\boxtimes)}$ and Mark H.M. Winands

Games and AI Group, Department of Data Science and Knowledge Engineering,
Maastricht University, Maastricht, The Netherlands
{c.sironi,m.winands}@maastrichtuniversity.nl

Abstract. General Game Playing (GGP) programs need a Game Description Language (GDL) reasoner to be able to interpret the game rules and search for the best actions to play in the game. One method for interpreting the game rules consists of translating the GDL game description into an alternative representation that the player can use to reason more efficiently on the game. The Propositional Network (PropNet) is an example of such method. The use of PropNets in GGP has become popular due to the fact that PropNets can speed up the reasoning process by several orders of magnitude compared to custom-made or Prolog-based GDL reasoners, improving the quality of the search for the best actions. This paper analyzes the performance of a PropNet-based reasoner and evaluates four different optimizations for the PropNet structure that can help further increase its reasoning speed in terms of visited game states per second.

1 Introduction

The aim of General Game Playing (GGP) is to develop programs that are able to play any arbitrary game at an expert level by being only given its rules. These programs must devise a playing strategy without having any prior knowledge about the game. Moreover, the rules are given to the player just before game playing starts and usually for each game step only few seconds are available to choose a move. Thus, the player has to learn an appropriate playing strategy on-line and in a limited amount of time.

To be able to play games, a GGP program has two main components: a way to interpret the game rules, written in the Game Description Language (GDL), and a strategy to choose which actions to play.

Regarding the first component, many different approaches have been proposed to parse the game rules. Three main methods to interpret GDL can be identified: (1) Prolog-based interpreters that translate the game rules from GDL into Prolog and then use a Prolog engine to reason about them, (2) custom-made interpreters written for the sole purpose of interpreting GDL rules, and (3) reasoners that translate the GDL description into an alternative representation that the player can use to efficiently reason about the game. A description and performance evaluation of available GDL reasoners is given in [7].

Regarding the second component, most of the approaches that proved successful in addressing the challenges of GGP are based on Monte-Carlo simulation

© Springer International Publishing AG 2017
T. Cazenave et al. (Eds.): CGW 2016/GIGA 2016, CCIS 705, pp. 133–151, 2017.
DOI: 10.1007/978-3-319-57969-6_10

techniques and especially on Monte-Carlo Tree Search (MCTS) [1,2]. For Monte-Carlo methods the choice of the best action to play is based on game statistics collected by sampling the state space of the game. The number of samples that Monte-Carlo methods can collect directly influences their performance. A higher number of samples in general improve the quality of the chosen actions.

A faster GDL reasoner, which in a given amount of time can analyze a higher number of game states than other reasoners, can positively influence Monte-Carlo based search. Propositional Networks (PropNets) [3,8] have become popular in GGP because they can speed up the reasoning process by several orders of magnitude compared to custom-made or Prolog-based GDL reasoners. Nowadays, all the best GGP programs use a PropNet-based reasoner [4,5,9].

The purpose of this paper is to analyze the performance of the implementation of the PropNet-based reasoner provided in the GGP-Base framework [9], discuss four optimizations of the structure of the PropNet and empirically evaluate their impact on the speed of the reasoning process. The performance of the custom-made GDL reasoner provided in the GGP-Base framework, called GGP-Base Prover, has been used as a reference.

The reminder of the paper is structured as follows. Section 2 gives a short introduction to GDL and PropNets. Sections 3 and 4 give some details about the PropNet implementation and a description of the PropNet optimizations respectively. Section 5 presents the empirical evaluation of the PropNet and Sect. 6 concludes and indicates potential future work.

2   Background

PropNets are one of the promising representations that can be used to reason about GDL descriptions. Subsection 2.1 gives a brief introduction to GDL and Subsect. 2.2 briefly describes the structure of a PropNet.

2.1   The Game Description Language

The Game Description Language (GDL) is a first order logic language used in GGP to represent the rules of games [6]. In GDL a game state is defined by specifying which propositions are true in that state. A set of reserved keywords is used to define the characteristics of the game.

Figure 1 shows as an example the GDL description of a simple game, where a player can independently turn on two lights (p and q). After being turned on, each light will remain on. The game ends when both lights are on and the player achieves a goal with score 100. In the figure, the GDL keywords are represented in bold.

2.2   The PropNet

A Propositional Network (PropNet) [3,8] can be seen as a graph representation of GDL. Each component in the PropNet represents either a proposition or a

```
(role player)
(light p) (light q)
(<= (legal player (turnOn ?x)) (not (true (on ?x))) (light ?x))
(<= (next (on ?x)) (does player (turnOn ?x)))
(<= (next (on ?x)) (true (on ?x)))
(<= terminal (true (on p)) (true (on q)))
(<= (goal player 100) (true (on p)) (true (on q)))
```

Fig. 1. Example of GDL game description.

logic gate. Propositions can be distinguished into three types: *input* propositions that have no input components, *base* propositions that have one single *transition* as input, and *view* propositions that are the remaining ones. The truth values of *base* propositions represent the state of the game. The dynamics of the game are represented by *transitions* that are identity gates that output their input value with one step delay and control the truth values of *base* propositions in the next step. The truth value of every other component is a function of the truth value of its inputs, except for *input* propositions, for which the game playing agent sets a value when choosing the action to play. Figure 2 shows as an example the PropNet that corresponds to the GDL description given in Fig. 1.

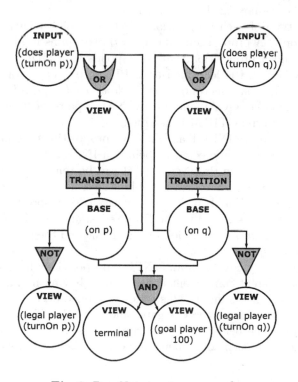

Fig. 2. PropNet structure example.

3   PropNet Implementation

To create the PropNet the algorithm provided in the GGP-Base framework
is used.[1] This algorithm is implemented in the *create(List< Gdl> description)*
method of the *OptimizingPropNetFactory* class and builds the PropNet according to the rules in the given GDL description.

The final product of the algorithm is a set of all the components in the Prop-
Net, each of which has been connected to its input and output components. This
set can then be used to initialize a PropNet object. The algorithm distinguishes
six different types of components: *constants* (TRUE and FALSE), *propositions*,
transitions and three different *gates* (AND, OR, NOT).

The GGP-Base framework also provides a PropNet class that can be initialized using the created set of components. We used this class as a starting point
and implemented some changes to the initialization process to ensure that the
PropNet respects certain constraints that are needed for the optimizations algorithms to work consistently. The first step of the initialization iterates over all
the components in the PropNet and inserts them in different lists according to
their type. While iterating over all the components, the following are the main
actions that the initialization algorithm performs:

- Identify a single TRUE and a single FALSE constant, creating them if they
  do not exist, or removing the redundant ones.
- Identify the type of each proposition. Each proposition must be associated
  to one type only. A proposition that has a *transition* as input is identified as
  BASE type and a proposition that corresponds to a GDL relation containing the *does* keyword is identified as INPUT type. The propositions corresponding to GDL relations containing the *legal, goal* or *terminal* keyword are
  identified as LEGAL, GOAL and TERMINAL type respectively. To all other
  propositions the type OTHER is assigned.
- Make sure that all the INPUT and LEGAL propositions are in a 1-to-1 relation. If a proposition is detected as being an INPUT but there is no corresponding LEGAL in the PropNet, then it can be removed since we are sure
  that the corresponding move will never be chosen by the player. On the contrary, if there is a LEGAL proposition with no corresponding INPUT, the
  INPUT proposition is added to the PropNet, since the LEGAL proposition
  might become true at a certain point of the game and the player might choose
  to play the corresponding move.
- Make sure that only constants and INPUT propositions have no input components. If a different component is detected as having no inputs, set one of
  the two constants as its input. This action is needed because as a by-product
  of the PropNet creation some OR gates and non-INPUT propositions might
  have no inputs. The behavior of the PropNet has been empirically tested to
  be consistent when such components are connected to the FALSE constant.

---

[1] We have used a more recent and improved version than the one tested in [7].

4 Optimizations

The PropNets built by the algorithm given in the GGP-Base framework [9] contain usually many components that are not strictly necessary to reason about the game. This section presents four optimizations that can be performed on the PropNet structure to reduce the number of these components. Opt0 (Subsect. 4.1) removes components that are known to have a constant truth value, Opt1 (Subsect. 4.2) removes propositions that do not have a particular meaning, Opt2 (Subsect. 4.3) detects more constant components and removes them, and Opt3 (Subsect. 4.4) removes components that have no output and are not influential. All the optimization algorithms except the last one are already provided in the GGP-Base framework. The algorithms described here contain some minor modifications with respect to the original GGP-Base version in order to adapt them to the changes that were performed on the PropNet class structure.

4.1 Opt0: Remove Constant-Value Components

This optimization removes from the PropNet the components that are known to be always *true* or always *false* and at the same time do not have a particular meaning for the game. For example an AND gate that has an input that is always *false* will also always output *false*, thus the gate can be removed and all its outputs can be connected directly to the *FALSE* constant of the PropNet.

Algorithm 1 shows the main steps of this optimization. The sets O_T and O_F, at any moment, contain respectively the outputs of the *TRUE* and the outputs of the *FALSE* constant that still have to be checked for removal. At the beginning O_T contains all the outputs of the *TRUE* constant and O_F contains all the outputs of the *FALSE* constant (Lines 2 and 3).

The procedure REMOVEFROMTRUE(*propnet*, O_T, O_F) (Line 5) and the procedure REMOVEFROMFALSE(*propnet*, O_T, O_F) (Line 6) check the outputs of the *TRUE* and of the *FALSE* constant respectively. Algorithm 2 shows exactly which components the first procedure removes. The algorithm for the second procedure removes the outputs of the FALSE constant in a similar way. In the case of the FALSE constant, also always false GOAL and LEGAL propositions are removed since they will never be used. Moreover, whenever a LEGAL proposition is removed also the corresponding INPUT proposition is removed, since it is certain that the corresponding move will never be played.

Note that whenever a component is removed or detected as having always a constant value, it means that also its output is constant, thus its output components are connected directly to one of the two constants. In this case each output component will be added to the appropriate set (either O_T or O_F) to be checked in the next steps.

Algorithm 1 alternates between the two procedures mentioned above until both sets, O_T and O_F, are empty. This repetition is needed because of the NOT gate. Whenever this gate is removed from the outputs of a constant, its outputs are connected to the other constant, thus the set of outputs to be checked for that constant will still have at least one element.

Algorithm 1. Remove constant-value components

1: **procedure** $\text{OPT}_0(propnet)$
2:     $O_T \leftarrow propnet.TRUE.outputs$
3:     $O_F \leftarrow propnet.FALSE.outputs$
4:     **while** $O_T \neq \emptyset$ **or** $O_F \neq \emptyset$ **do**
5:         $\text{REMOVEFROMTRUE}(propnet, O_T, O_F)$
6:         $\text{REMOVEFROMFALSE}(propnet, O_T, O_F)$
7:     **end while**
8: **end procedure**

Algorithm 2. Remove true components

1: **procedure** $\text{REMOVEFROMTRUE}(propnet, O_T, O_F)$
2:     **while** $O_T \neq \emptyset$ **do**
3:         $c \leftarrow O_T.removeElement()$
4:         **switch** $c.compType$ **do**
5:             **case** TRANSITION
6:                 **if** $|c.outputs| = 0$ **then**
7:                     $propnet.remove(c)$
8:                 **end if**
9:             **case** NOT
10:                 connect $c.outputs$ to FALSE
11:                 $O_F \leftarrow O_F \cup c.outputs$
12:                 $propnet.remove(c)$
13:             **case** AND
14:                 **if** $|c.inputs| = 1$ **then**   ▷ Only TRUE as input
15:                     connect $c.outputs$ to TRUE
16:                     $O_T \leftarrow O_T \cup c.outputs$
17:                     $propnet.remove(c)$
18:                 **else if** $|c.inputs| = 2$ **then**   ▷ Only 2 inputs, one is TRUE
19:                     connect $c.outputs$ to other input
20:                     $propnet.remove(c)$
21:                 **else**   ▷ More than 2 inputs, one is TRUE
22:                     disconnect c form TRUE
23:                 **end if**
24:             **case** OR
25:                 connect $c.outputs$ to TRUE
26:                 $O_T \leftarrow O_T \cup c.outputs$
27:                 $propnet.remove(c)$
28:             **case** PROPOSITION
29:                 connect $coutputs$ to TRUE
30:                 $O_T \leftarrow O_T \cup c.outputs$
31:                 **if** $c.propType \in \{OTHER, BASE\}$ **then**
32:                     $propnet.remove(c)$
33:                 **end if**
34:         **end switch**
35:     **end while**
36: **end procedure**

4.2   Opt1: Remove Anonymous Propositions

This optimization is trivial, nevertheless useful as it removes many useless components from the PropNet. The algorithm for this optimization (Algorithm 3) simply iterates over all the propositions in the PropNet and removes the ones with type OTHER, connecting their input directly to each of their outputs. These propositions can be safely removed as they do not have any special meaning for the game.

Algorithm 3. Remove anonymous propositions

1: **procedure** OPT1(*propnet*)
2:     **for all** $p \in propnet.propositions$ **do**
3:         **if** $p.propType = $ OTHER **then**
4:             connect $p.input$ with $p.outputs$
5:             $propnet.remove(p)$
6:         **end if**
7:     **end for**
8: **end procedure**

4.3   Opt2: Detect and Remove Constant-Value Components

This optimization can be seen as an extension of Opt0 where, before removing from the PropNet the constant value components directly connected to the *TRUE* and *FALSE* constant, the algorithm detects if there are other constant value components that have not been discovered yet.

This optimization (see Algorithm 4) associates to each component c in the PropNet a set V_c that contains all the truth values that such component can assume during the whole game. There are only four possible sets of truth values, namely:

– $N = \emptyset$: if the corresponding component can assume *neither* of the truth values.
– $T = \{true\}$: if the corresponding component can only be *true* during all the game.
– $F = \{false\}$: if the corresponding component can only be *false* during all the game.
– $B = \{true, false\}$: if the corresponding component can assume *both* values during the game.

The idea behind the algorithm is to start from the components for which the truth value that they will assume in the initial state of the game is known. It then propagates this value to each of their outputs o and updates the corresponding truth value set V_o. Whenever the truth values set of a component is updated, the algorithm propagates such changes on to its output components. This process

---

Algorithm 4. Detect and remove constant-value components

1: **procedure** OPT2(*propnet*)
2:     Initialize all the parameters and the stack S
3:     **while** $S \neq \emptyset$ **do**
4:         $(c, P_i) \leftarrow S.pop()$
5:         $O_c \leftarrow$ TOOUTPUTVALUESET(c, P_i)
6:         $P_c \leftarrow O_c \setminus V_c$
7:         **if** $P_c \neq N$ **then**
8:             $V_c \leftarrow V_c \cup P_c$
9:             **for all** $o \in c.outputs$ **do**
10:                 $S.push(o, P_c)$
11:             **end for**
12:             **if** $c.compType =$ PROPOSITION **and** $c.propType =$ LEGAL **then**
13:                 $i \leftarrow c.correspondingInput$
14:                 $S.push(i, P_c)$
15:             **end if**
16:         **end if**
17:     **end while**
18:     **for all** $c \in propnet.components$ **do**
19:         **if** $V_c = T$ **or** $V_c = F$ **then**
20:             Connect c to the appropriate constant
21:         **end if**
22:     **end for**
23:     OPT0(*propnet*)
24: **end procedure**

---

will eventually end when the truth values sets of all components stop changing. Termination is guaranteed since only the truth values just added to the truth values set of a component are propagated to its outputs and the number of possible truth values is finite.

When the algorithm starts, the set V_c of each component c is set to N, since it is not known yet which values the component can assume. For each AND gate a the algorithm keeps track of TI_a, i.e. the number of inputs of a that can assume the *true* value. Similarly, for each OR gate o the algorithm keeps track of FI_o, i.e. the number of inputs of o that can assume the *false* value. This parameters are used to detect when an AND gate and an OR gate can assume respectively the *true* (if $TI_a = |a.inputs|$) and the *false* (if $FI_o = |o.inputs|$) value. These values are initialized to 0 for all the gates.

The algorithm exploits a stack structure S to keep track of the components for which the set of truth values that their input(s) can assume is changed. A pair (c, P_i) is added to the stack when the algorithm detects that an input i of the component c can also assume the values in the set $P_i \subseteq V_i$, and such values must be propagated to the component c. At the beginning the stack is filled with the following pairs:

– $(TRUE, T)$, the $TRUE$ constant can assume value *true*.
– $(FALSE, F)$, the $FALSE$ constant can assume value *false*.

- (i, F), for each INPUT proposition i in the PropNet. Each INPUT proposition can be *false* since we assume that no game exists where one player can only play a single move for the whole game.
- (b_j, T), for each BASE proposition b_j in the PropNet that is *true* in the initial state.
- (b_j, F), for each BASE proposition b_j in the PropNet that is *false* in the initial state.

During each iteration, the algorithm pops a pair (c, P_i) from the stack (Line 4) and checks if, given the new truth values P_i that the input i can assume, also the truth values V_c of its output c will change. Note that not for each type of component the set of truth values that its input can assume corresponds to the set of truth values that the component itself can output. The NOT component n, for example, has $V_n = T$ if its input i has $V_i = F$. Moreover, for an AND gate a, $true \in V_a \Leftrightarrow true \in V_i, \forall i \in a.inputs$. The same holds for the *false* value for an OR gate. This means that the algorithm must first change the values in P_i according to the type of the component c, obtaining the new set of truth values O_c that c can output. This is done at Line 5 by the function TOOUTPUTVALUESET(c, P_i). Subsequently, the algorithm checks if in O_c there are some values P_c that were not in V_c yet (Line 6), and if so, it adds them to the set V_c (Line 8) and records on the stack that they have to be propagated to all the outputs o of c (Lines 9–11). Here the algorithm treats each LEGAL propositions as if it was a direct input of the corresponding INPUT proposition, thus whenever the truth values set of a LEGAL proposition changes, the values are propagated to the corresponding INPUT proposition (Lines 12–15).

When no more changes are detected in the truth values sets (Line 3), the process terminates. At this point, the truth values set of each component is checked (Line 19) and if it equals the set T or F it is certain that the component will always be respectively *true* or *false*. It can then be disconnected from its input(s) and connected to the correct constant (Line 20).

The last step the algorithm performs consists in running the same algorithm that was proposed as Opt0 to remove all the newly detected constant components (Line 23).

4.4 Opt3: Remove Output-Less Components

This optimization is also quite trivial, but helps remove some more useless components. Algorithm 5 shows this procedure: all the components in the PropNet are checked, if they are gates, or propositions of type OTHER and they have no output they are removed from the PropNet. Every time a component is removed, its inputs are added again to the set of components to be checked, since removing their outputs might have made them output-less.

5 Empirical Evaluation

In this section an empirical evaluation of the performance of the PropNet and its optimizations is presented. Subsection 5.1 describes the setup of the

---

Algorithm 5. Remove output-less components

1: **procedure** OPT3(*propnet*)
2:     $Q \leftarrow propnet.components$
3:     **while** $Q \neq \emptyset$ **do**
4:         $c \leftarrow Q.removeElement()$
5:         **if** ((*c.compType* = PROPOSITION **and** *c.propType* = OTHER)
            or *c.compType* $\in$ {AND, OR, NOT}) **and** $|c.outputs| = 0$ **then**
6:             $Q \leftarrow Q \cup c.inputs$
7:             *propnet.remove(c)*
8:         **end if**
9:     **end while**
10: **end procedure**

---

performed experiments. Subsections 5.2 and 5.3 discuss the results of the experiments that compare the performance of single optimizations and combinations of them respectively. The combination of PropNet optimizations that performs overall best is then compared with the default Prover. Subsection 5.4 presents a comparison of PropNet and Prover in terms of their speed, while Subsect. 5.5 presents a comparison in terms of their game-playing performance.

5.1   Setup

To evaluate the performance of the PropNet multiple series of experiments are performed. Each of them tests the performance of the PropNet with different optimizations and combinations of them. Each series of experiments poses the bases to decide which other combinations of optimizations to check.

The different PropNet optimizations and their combinations are tested using flat Monte-Carlo Search (MCS) on a set of heterogeneous games. For each optimized PropNet the search is run from the initial state of the game with a time limit of 20 s. This experiment is repeated 100 times for each of the chosen games. Such games are the following: *Amazons, Battle, Breakthrough, Chinese Checkers* with 1, 2, 3, 4 and 6 players, *Connect 4, Othello, Pentago, Skirmish* and *Tic Tac Toe*. The GDL descriptions of these games can be found on the GGP-Base repository [10].[2]

One of the reasons behind the choice of repeating each experiment multiple times for each game is that for each repetition of the game a different seed is used for the random number generator that controls the random exploration of the search tree with the MCS algorithm. Thus, for different seeds different results might be obtained and different parts of the search space explored.

Another reason is that the number of components that the PropNet of a game has when created by the basic algorithm (i.e. without optimizations) is not always constant. This variance in the number of components could be due

---

[2] The GDL descriptions used for the experiments were downloaded from the repository on 03/02/2016.

to the non-determinism of the order in which game rules are translated into Prop-Net components for different runs of the algorithm. This can cause a different grounding order of the GDL description, originating more or less propositions and can also cause gates and propositions to be connected in different equivalent orders.

The optimized PropNet that showed the best overall performance in the previous series of experiments is compared with the GGP-Base Prover in another series of experiments. Both reasoners are also tested with the addition of a cache that memorizes the queries results.

This series of experiments matches two MCS-based players that use the Prover, one with cache and one without, against each other, and two MCS-based players that use the best optimized PropNet, one with cache and one without, against each other. We use the same 13 games that were used for the other experiments. Each player has 10 s per move to perform the search. A new PropNet is built for each match in advance, before the game playing starts. For each game, if r is the number of roles in the game, there are 2^r different ways in which 2 types of players can be assigned to the roles [11]. Two of the configurations involve only the same player type assigned to all the roles, thus are not interesting and excluded from the experiments. Each configuration is run the same number of times until at least 100 games have been played in total.

At the end of each game repetition the speed of the reasoners is computed by dividing the total number of nodes visited by the total time spent on the search during the whole game. Since we are only interested in the reasoning speed, for this experiment we do not consider the 10 s search time per move strictly, but we allow each player to finish the current simulation when this time expires.

The final series of experiments aims at evaluating the impact of the reasoners on the win rate of game playing agents. This experiments match two MCTS-based players, one that uses the fastest version of the Prover (i.e. with the cache) and one that uses the fastest optimized PropNet (also with the cache), against each other. The settings are the same as in the previous experiment, except the minimum number of played games that is increased to 200. Moreover, for this experiment the 10 s search time per move is considered strictly.

Before running any of the described experiments, the PropNet and all its optimized versions were tested against the Prover for consistency. For each game (about 300) in the GGP-Base repository [10], for a duration of 60 s, the same random simulations were performed querying both the Prover and the currently tested version of the PropNet for next states, legal moves, terminality and goals in terminal states. The results returned by the PropNet were compared with the ones returned by the Prover for consistency. All the PropNet versions passed this test on all the games in the repository, except for 12 games for which the PropNet construction could not be completed in the given time.

In all experiments, a limit of 10 min was given to the program to build the PropNet. The experiments that compare the speed of PropNet and Prover with and without cache were performed on an AMD Opteron 6174 2.2-GHz. All other experiments were performed on an AMD Opteron 6274 2.2-GHz.

5.2   Comparison of Single Optimizations

The first series of experiments compares with the basic version of the PropNet (BasicPN) the performance of each of the previously described optimizations applied singularly (Opt0, Opt1, Opt2, Opt3). Table 1 shows the obtained results. For each PropNet variant, for each game the first block of the table gives the average simulation speed in nodes per second, the second block gives the average number of components and the third block gives the average total initialization time (creation+optimization+state initialization) in milliseconds. The line at the bottom of each block reports the average over the 13 games of the percentage increase of the values considered in the block, relative to the basic version of the PropNet (BasicPN).

The main interest is the speed increase that the optimizations induce on the PropNet, however the other two aspects are also relevant. A low number of components means less memory usage, and a shorter initialization time means more time for metagaming at the beginning of a match (or more chances to avoid timing out when the start clock time is short). From the table it seems that for most of the games, as expected, the increase in the simulation speed is related to the decrease in the number of components in the PropNet.

As can be seen, none of the optimizations outperforms the others in speed for all games. Opt0 and Opt2 seem to have the best performance in *Amazons*, *Battle*, *Othello* and *Connect 4*, while Opt1 performs best in the other games. When looking at the initialization time, Opt2 is the one that increases it the most for almost all the games. Another observation is that the performance of Opt2 is overall better than the one of Opt0. This was expected because Opt2 is an extension of Opt0, thus for the same PropNet it always removes at least the same number of components as Opt0.

The speed is used as main criterion to choose which of the four optimization to use as starting point for further experiments that involve testing combinations of optimizations. If we consider the speed, Opt0 and Opt2 are the ones that, on average, produce the highest increase. However, the high average is due to the considerable relative increase that they produce in *Othello*. If we consider the optimization that produces the highest speed in most of the games, then Opt1 is the most suitable to be selected. Moreover, Opt1 is the optimization that reduces the most the number of components of the PropNet without consistently slowing down the initialization process.

5.3   Comparison of Combined Optimizations

In this series of experiments Opt1 is combined with other optimizations applied in sequence. In general, when we refer to OptXY we refer to the PropNet optimization obtained by applying OptX and OptY in sequence. These experiments first compare the combinations of optimizations Opt13, Opt12 and Opt102. The combination Opt10 has been excluded from the test since it is considered less interesting. As also previously mentioned, Opt0 always removes a subset of the components that are removed by Opt2, thus Opt10 is expected to perform less

Table 1. Comparison of single optimizations

	Game	BasicPN	Opt0	Opt1	Opt2	Opt3
Avg. speed (nodes/second)	Amazons	35.1	41.4	32.7	41	40.2
	Battle	34957	49666	37877	51257	35276
	Breakthrough	50557	50932	65518	51357	51058
	Chinese Checkers 1P	426374	427773	550230	444671	424516
	Chinese Checkers 2P	125581	128623	189368	128910	127519
	Chinese Checkers 3P	155886	157242	169352	161000	159267
	Chinese Checkers 4P	105766	106738	127886	107153	105660
	Chinese Checkers 6P	119650	118547	126863	113700	118783
	Connect 4	110081	113484	105081	112920	109672
	Othello	290	1610	235	1604	295
	Pentago	76336	76786	116065	76721	96782
	Skirmish	5887	6022	6780	6230	6151
	Tic Tac Toe	223403	228056	248769	234915	222952
	Avg. relative increase	–	40.59%	15.51%	41.44%	3.95%
Avg. number of components	Amazons	1497649	1254742	741874	1192364	1023913
	Battle	51197	14267	36863	14262	50721
	Breakthrough	10745	10678	5933	10678	10584
	Chinese Checkers 1P	793	785	559	785	789
	Chinese Checkers 2P	1540	1524	1179	1524	1532
	Chinese Checkers 3P	2411	2389	1845	2236	2400
	Chinese Checkers 4P	3159	3119	2465	2999	3133
	Chinese Checkers 6P	4451	4411	3473	4123	4431
	Connect 4	2164	2063	1724	1291	2114
	Othello	1311988	274940	1033197	274940	1305515
	Pentago	3696	3706	1470	3708	2111
	Skirmish	126019	124267	108171	124267	78575
	Tic Tac Toe	312	291	249	291	302
	Avg. relative increase	–	−14.28%	−29.21%	−18.62%	−9.49%
Avg. total init. time (ms)	Amazons	311335	313719	314455	417097	315637
	Battle	5756	6027	5897	6303	5869
	Breakthrough	3989	4007	4012	4358	3910
	Chinese Checkers 1	2699	2651	2659	2653	2707
	Chinese Checkers 2	2848	2773	2810	2873	2775
	Chinese Checkers 3	3162	3140	3159	3251	3149
	Chinese Checkers 4	3258	3261	3241	3473	3244
	Chinese Checkers 6	3225	3203	3204	3639	3205
	Connect 4	2437	2465	2456	2698	2430
	Othello	35756	36486	37074	39417	36544
	Pentago	4249	4230	4278	4390	4232
	Skirmish	11887	11702	11664	12089	11824
	Tic Tac Toe	1525	1529	1523	1522	1508
	Avg. relative increase	–	0.13%	0.24%	7.69%	−0.19%

than Opt12. However, Opt0 has less negative impact than Opt2 on the total initialization time. This is why these experiments include the test of Opt102: we want to see if the application of Opt0 before Opt2 can speed up the process of Opt2 that will then run on a smaller PropNet.

The results of this series of experiments can be seen in columns 3, 4 and 5 of Table 2. The structure of this table is the same as Table 1. The average percentage increase reported in the last line of each block is still computed with respect to the basic version of the PropNet (BasicPN).

As the table shows, regarding the speed, Opt12 seems to be the one achieving the best overall performance. However, the performance of Opt102 is rather close, as expected, because these two combinations should reduce each PropNet to the same number of components. The small difference in performance is probably due the reasons already mentioned in Sect. 5.1. Both the difference in the random seed used for each repetition of the game and the variance in the number of components generated by the algorithm that creates the initial PropNet can influence the performance.

One more thing that can be noticed from Table 2 is that running Opt0 before Opt2 helps reducing the initialization time for large games, while it seems to have almost no effect on smaller games. Moreover, Opt13 is the one that, regarding the speed, performs worse in this series of experiments, thus it has been excluded from further tests. Among Opt12 and Opt102, it has been chosen to keep testing on top of Opt102 because of its shorter initialization time for games with large PropNets, given that its speed is still comparable with the one of Opt12.

Using Opt102 as starting point, there is only one more interesting combination of optimizations left to test: Opt1023. No further gain in performance can be obtained by repeating the same optimizations multiple times in a row, since no further change will take place in the structure of the PropNet. Thus, it is not interesting to evaluate combinations of optimizations that extend Opt1023.

The last column of Table 2 shows the statistics for Opt1023. For most of the games, Opt1023 seems to be the fastest. It is also the one that reduces the number of PropNet components the most. As for the initialization time, this optimization is between a few milliseconds and a bit more than 1 second slower that the basic version of the PropNet, except for *Amazons*. Optimizing the large PropNet of *Amazons* can slow down the initialization time by more than a minute.

5.4   Comparison of PropNet and Prover

In this series of experiments the overall fastest combination of optimizations among the tested ones (Opt1023) is compared with the Prover. More precisely, Opt1023 and the Prover are compared measuring their speed over complete games (as opposed to previous experiments that were comparing the speed only on the first step of the game).

Moreover, for both of them also a cached version is tested (i.e. CachedProver and CachedOpt1023). The GGP-Base framework [9] provides a cache structure

Table 2. Comparison of combined optimizations

	Game	BasicPN	Opt12	Opt102	Opt13	Opt1023
Avg. speed (nodes/second)	Amazons	35	38.5	41.4	32.3	41
	Battle	34957	59308	59697	39981	60419
	Breakthrough	50557	66943	66551	66833	66991
	Chinese Checkers 1P	426374	570858	562737	541682	561634
	Chinese Checkers 2P	125581	194442	192048	190161	193752
	Chinese Checkers 3P	155886	175410	176162	170722	176185
	Chinese Checkers 4P	105766	130362	130279	129194	130451
	Chinese Checkers 6P	119650	127535	128111	127619	129000
	Connect 4	110081	127053	126535	105978	129272
	Othello	290	1934	1894	245	1979
	Pentago	76336	116353	115064	117127	121108
	Skirmish	5887	7075	7042	7403	7600
	Tic Tac Toe	223403	259980	257285	247246	257525
	Avg. relative increase	–	70.32%	69.39%	17.38%	73.48%
Avg. number of components	Amazons	1497649	623460	623460	711596	596240
	Battle	51197	11084	11077	36676	10902
	Breakthrough	10745	5900	5900	5869	5836
	Chinese Checkers 1P	793	556	556	559	556
	Chinese Checkers 2P	1540	1172	1172	1179	1172
	Chinese Checkers 3P	2411	1718	1718	1845	1718
	Chinese Checkers 4P	3159	2362	2362	2465	2362
	Chinese Checkers 6P	4451	3238	3238	3473	3238
	Connect 4	2164	1063	1063	1724	1056
	Othello	1311988	208510	208510	1031580	206846
	Pentago	3696	1464	1473	1338	1337
	Skirmish	126019	107296	107296	62427	61552
	Tic Tac Toe	312	239	239	249	239
	Avg. relative increase	–	–42.34%	–42.32%	–32.52%	–45.65%
Avg. total init. time (ms)	Amazons	311335	411905	400113	312793	401559
	Battle	5756	6367	6233	5968	6329
	Breakthrough	3989	4354	4415	3982	4328
	Chinese Checkers 1P	2699	2693	2654	2707	2652
	Chinese Checkers 2P	2848	2848	2843	2817	2842
	Chinese Checkers 3P	3162	3214	3186	3160	3167
	Chinese Checkers 4P	3258	3405	3330	3275	3379
	Chinese Checkers 6P	3225	3423	3430	3207	3395
	Connect 4	2437	2536	2555	2417	2525
	Othello	35756	39170	36689	35359	37804
	Pentago	4249	4269	4286	4308	4325
	Skirmish	11887	12386	12285	11870	12577
	Tic Tac Toe	1525	1532	1535	1524	1555
	Avg. relative increase	–	6.38%	5.18%	0.18%	5.66%

Table 3. Comparison of the PropNet with the Prover and effect of the cache

	Game	Prover	CacheProver	Opt1023	CacheOpt1023
Avg. speed (nodes/second)	Amazons	5.7	2316	28.1	30519
	Battle	45.2	2457	38656	36607
	Breakthrough	235	241	56275	51569
	Chinese Checkers 1P	2273	466014	532426	862408
	Chinese Checkers 2P	1478	93251	159935	258639
	Chinese Checkers 3P	1105	28300	118160	133733
	Chinese Checkers 4P	536	32684	82955	117017
	Chinese Checkers 6P	607	5744	57008	53230
	Connect4	180	2455	122325	207508
	Othello	3.2	5502	649	80328
	Pentago	152	155	93185	75998
	Skirmish	26	4081	2997	3946
	Tic Tac Toe	1650	287380	225127	547398
	Avg. relative increase	–	22139%	–	9321%

that memorizes the results returned by the underlying reasoner and prevents it from computing the same queries multiple times.

The results of these experiments are shown in Table 3. The last row of this table reports for both CachedProver and CachedOpt1023 the average percentage increase of the speed with respect to their non-cached versions.

From the table it is visible how the optimized PropNet achieves a much better performance than the Prover in the considered games. When adding the cache to both reasoners the difference in performance is reduced, however the PropNet is still faster in all games but one, *Skirmish*, for which the speed of the cached PropNet and the cached Prover are quite close.

The use of a cache provides some benefits increasing the overall performance of both reasoners with respect to their non-cached version. However, the cache gives more benefits to the Prover. For the Prover the speed is increased for all the games, while for the PropNet it is increased for most, but not all of them. To be noticed is that the increase in speed provided by the cache is especially relevant in the games of *Amazons* and *Othello*.

Moreover, observing the results for all the *Chinese Checkers* versions it is clear that the speed of the cached Prover and the speed of the cached PropNet both decrease when increasing the number of players. However, for the PropNet this decrease is slower. For *Chinese Checkers* with 1 player the cached PropNet is about 2 times faster than the cached Prover, while for the version with 6 players it is almost 10 times faster.

When performing the experiments it was also noticed that in many games the cache decreases the speed of the PropNet reasoner during the initial steps. This loss is then balanced towards the endgame, when the chance of finding cached query results increases. It takes some time for the cache to be filled with a sufficient number of entries and thus have a positive impact on the speed of the PropNet.

The same effect was not observed for the Prover. For the first steps of the games the cache did not decrease the speed of the Prover for any of the games, and for some of them increased it. The explanation for this is that the time for computing the answer of a query with the Prover is in general much higher than the one of the PropNet. Thus, for the Prover finding in the cache even a small number of query results saves enough computational time to compensate the extra time spent looking in the cache for results that are not present yet.

Finally, the results of Table 3 also help putting the PropNet into perspective with the other GDL reasoners analyzed in the paper [7]. Even if that paper uses different experimental settings than ours, we can still make some general observations. Considering the performance of the reasoners that, like the PropNet, rely on an alternative representation of the GDL description, it seems that our implementation of the PropNet provides for most of the games a speed increase of the same order of magnitude when compared to the Prover. Moreover, for *Amazons*, *Othello* and *Chinese Checkers* with 4 and 6 players, it seems that our optimized PropNet, especially with the cache, could even achieve a better performance in similar circumstances.

5.5   Game Playing Performance

In this series of experiments an MCTS player that uses the cached PropNet reasoner with the fastest combination of optimizations (Opt1023) is matched against an MCTS player that uses the cached Prover. Because Sect. 5.4 showed the cache to be overall beneficial for both reasoners, it has been included in this experiment.

Table 4 shows the win percentage of the cached PropNet-player against the cached Prover-player. The table does not include the results for the single-player version of *Chinese Checkers* because this game is tested separately and the score is used to measure the performance of the players. This game has a relatively small search space, so both players achieved the maximum score in every match.

Moreover, no results are shown for *Amazons* and *Othello* because for both games, during the first game steps, the cached Prover-player could not return a move within the given time limit. Even with the use of the cache, during the first game steps the number of memorized query results is not sufficient to allow the Prover to complete even one MCTS simulation within the time limit.

Looking at the results for the remaining games, for most of them the cached PropNet-player achieves a win percentage close or equal to 100%. The games in which the performance of the cached PropNet-player seems to drop are the ones with more than 2 players. *Chinese Checkers* with 4 and 6 players are the ones where the win percentage for the cached PropNet-player is the lowest, but it is still significantly better than the one of the cached Prover-player. The game of *Tic Tac Toe* is the only exception, because its state space is so small that both

Table 4. Win percentage of the PropNet-player against the Prover-player

Game	Opt1023
Battle	100.0($\pm$0.0)
Breakthrough	100.0($\pm$0.0)
Chinese Checkers 2P	96.0($\pm$2.72)
Chinese Checkers 3P	77.5($\pm$5.75)
Chinese Checkers 4P	68.1($\pm$6.32)
Chinese Checkers 6P	64.7($\pm$5.73)
Connect 4	99.3($\pm$1.09)
Pentago	100.0($\pm$0.0)
Skirmish	100.0($\pm$0.0)
Tic Tac Toe	50.0($\pm$0.0)

players can easily reach a sufficient number of simulations to play optimally and result in a tie.

The results of Table 4 are in line to what would be expected when looking at the average speed reported in Table 3 for the two cached reasoners. For all the games for which the speed of the cached PropNet is at least one order of magnitude higher than the one of the cached Prover, the cached PropNet-player achieves a significantly higher win percentage. However, for the game of *Skirmish* Table 3 reports a similar average speed for both the cached reasoners so their performance would be expected to be close. A win rate of 100% for the cached PropNet can be explained by the fact that the speed per game step of the Prover exhibits a higher variance than the speed of the PropNet. The speed of the PropNet in the initial game steps is close to the average speed. The Prover, instead, is about 340 times slower than the PropNet in this stage. Its speed increases only in the last few steps of the game. At this point the PropNet-player already gained enough advantage over the Prover-player to win the game.

6 Conclusion and Future Work

In this paper the performance of a PropNet-based reasoner has been evaluated, together with four possible optimizations of the structure of the PropNet and their impact on the performance. Even though the tested implementation of the PropNet is based on the code provided by the GGP-Base framework, the principles behind its representation and its optimizations can also be applied in general.

Experiments have shown that the use of a PropNet substantially increases the reasoning speed by, on average, at least two orders of magnitude with respect to the GGP-Base Prover. Moreover, the addition of a combination of optimizations that reduce the size of the PropNet increases the reasoning speed further.

Experiments also show that the reasoning speed increase has a positive effect on the performance of the PropNet-based player. This player achieves a win rate close to 100% in most of the games for which it is matched against an equivalent player based on the Prover. Thus, it is possible to conclude that for a general game playing agent a reasoner based on a PropNet, especially when optimized, is in general a better choice than a custom-made GDL interpreter like the Prover.

Also the use of a cache proved to be useful for the PropNet in most of the games. For small games its effect is already visible in the first steps, while for most of the other games it helps only during later game steps. However, we may conclude that the use of a cache is overall positive for a PropNet reasoner.

Future work could further investigate the use of the cache with the PropNet, for example by devising a strategy to detect for each game if and when the use of a cache is helpful. Finally, another interesting aspect that future work could consider is the impact that the use of different strategies to propagate truth values among the components of the PropNet would have on the reasoning speed.

Acknowledgments. This work is funded by the Netherlands Organisation for Scientific Research (NWO) in the framework of the project GoGeneral, grant number 612.001.121.

References

1. Björnsson, Y., Finnsson, H.: CadiaPlayer: a simulation-based general game player. IEEE Trans. Comput. Intell. AI Games **1**(1), 4–15 (2009)
2. Coulom, R.: Efficient selectivity and backup operators in Monte-Carlo tree search. In: van den Herik, H.J., Ciancarini, P., Donkers, H.H.L.M. (eds.) CG 2006. LNCS, vol. 4630, pp. 72–83. Springer, Heidelberg (2007). doi:10.1007/978-3-540-75538-8_7
3. Cox, E., Schkufza, E., Madsen, R., Genesereth, M.R.: Factoring general games using propositional automata. In: Björnsson, Y., Stone, P., Thielscher, M. (eds.) Proceedings of the IJCAI Workshop on General Intelligence in Game-Playing Agents (GIGA), pp. 13–20 (2009)
4. Draper, S., Rose, A.: Sancho GGP player (2014). http://sanchoggp.blogspot.nl/2014/07/sancho-is-ggp-champion-2014.html
5. Emslie, R.: Galvanise (2015). https://github.com/richemslie/galvanise
6. Love, N., Hinrichs, T., Haley, D., Schkufza, E., Genesereth, M.R.: General game playing: game description language specification. Technical report, Stanford University, Stanford, CA, USA (2008)
7. Schiffel, S., Björnsson, Y.: Efficiency of GDL reasoners. IEEE Trans. Comput. Intell. AI Games **6**(4), 343–354 (2014)
8. Schkufza, E., Love, N., Genesereth, M.: Propositional automata and cell automata: representational frameworks for discrete dynamic systems. In: Wobcke, W., Zhang, M. (eds.) AI 2008. LNCS (LNAI), vol. 5360, pp. 56–66. Springer, Heidelberg (2008). doi:10.1007/978-3-540-89378-3_6
9. Schreiber, S.: The general game playing base package (2013). https://github.com/ggp-org/ggp-base
10. Schreiber, S.: Games - base repository (2016). http://games.ggp.org/base/
11. Sturtevant, N.R.: An analysis of UCT in multi-player games. ICGA J. **31**(4), 195–208 (2008)

Grounding GDL Game Descriptions

Stephan Schiffel[✉]

School of Computer Science/CADIA,
Reykjavik University, Reykjavik, Iceland
stephans@ru.is

Abstract. Many state-of-the-art general game playing systems rely on a ground (propositional) representation of the game rules. We propose a theoretically well-founded approach using efficient off-the-shelf systems for grounding game descriptions given in the game description language (GDL).

1 Introduction

Games in General Game Playing are generally described in the game description language (GDL) [11]. While allowing to describe a large class of games and being theoretically well-founded, reasoning with GDL is generally slow compared to game specific representations [16]. This limits the speed of search, both for heuristic search methods, such as Minimax, as well as for simulation-based approaches, such as Monte Carlo Tree Search. Thus, an important aspect of General Game Playing is to find a better representation of the rules of a game that facilitates both fast search in the game tree as well as efficient meta-gaming analysis. Propositional networks [19] and binary decision diagrams (BDDs) [5] have been proposed for faster reasoning with the game rules. Both approaches, require that game descriptions be grounded, that is, translated into a propositional representation. Other meta-gaming approaches could also benefit from having a propositional description of the game rules as input. Some examples are finding symmetries in games [14], discovering heuristics for games (e.g., [12]), proving game properties [8,17] and factoring games [3,4].

The GGP-Base framework [20], which is the basis for a number of general game players, contains code for generating a propositional network [19] representing the game rules. This code requires computing ground instances of all rules in the game description. However, the code seems ad-hoc and it is not obvious whether and for which class of game descriptions it maintains the semantics of the game rules. With this paper, we propose to transform the game description into an answer set program [2] and use the grounder of a state-of-the-art answer set solving system to compute a propositional representation of the game. This has the advantage of using a highly optimized and well-tested system, as well as being theoretically well-founded. Our system also turns out to be able to handle more games than the GGP-Base framework and being significantly faster for many games.

© Springer International Publishing AG 2017
T. Cazenave et al. (Eds.): CGW 2016/GIGA 2016, CCIS 705, pp. 152–164, 2017.
DOI: 10.1007/978-3-319-57969-6_11

2   Related Work

In [9], the authors report two methods for grounding GDL, one using Prolog and another using dependency graphs. Both methods have some deficiencies and the authors only manage to ground 96 out of the 171 tested games. While the authors do not report on the size of the grounded game descriptions, they report on one of their methods to produce game descriptions that are unnecessarily big.

In [8], The authors use Answer Set Programming (ASP) [2] to prove properties of games by transforming a GDL description into an answer set program and adding constraints that encode the properties to be proven. The system they have implemented uses the Potassco ASP solver [6] which relies on grounding the answer set program, and thus, indirectly the game description.

3   Game Description Language (GDL)

The game description language [11, 18] is a first-order-logic based language that can be seen as an extension of Datalog permitting negations and function symbols. Thus, a game description in GDL is a logic program. The game specific semantics of GDL stems from the use of certain special relations, such as for describing the initial game state (init), detecting (terminal) and scoring (goal) terminal states, for generating legal moves (legal) and successor states (next). A game state is represented by the set of terms that are true in the state (e.g., cell(1,1,b)) and the special relations true(f) and does(r, m) can be used to refer to the truth of f being in the current state and role r doing move m in the current state transition. Figure 1 shows a partial GDL description for the game Tic-Tac-Toe.

GDL allows to describe a wide range of deterministic perfect-information simultaneous-move games with arbitrary number of adversaries. Turn-based games are modeled by having the players that do not have a turn return a move with no effect (e.g., noop in Fig. 1).

To ensure an unambiguous declarative interpretation, valid GDL descriptions need to fulfill a number of restrictions:

Definition 1. *The dependency graph for a set G of clauses is a directed, labeled graph whose nodes are the predicate symbols that occur in G and where there is a positive edge $p \xrightarrow{+} q$ if G contains a clause $p(\overline{s}) \Leftarrow \ldots \wedge q(\overline{t}) \wedge \ldots$, and a negative edge $p \xrightarrow{-} q$ if G contains a clause $p(\overline{s}) \Leftarrow \ldots \wedge \neg q(\overline{t}) \wedge \ldots$.*

To constitute a valid GDL specification, a set of clauses G and its dependency graph Γ must satisfy the following.

1. *There are no cycles involving a negative edge in Γ (this is also known as being stratified [1, 7]);*
2. *Each variable in a clause occurs in at least one positive atom in the body (this is also known as being allowed [10]);*

```
1  role(xplayer).
2  role(oplayer).
3
4  init(cell(1, 1, b)).
5  init(cell(1, 2, b)).
6  ...
7  init(cell(3, 2, b)).
8  init(cell(3, 3, b)).
9  init(control(xplayer)).
10
11 legal(W, mark(X, Y)) :-
12    true(cell(X, Y, b)),
13    true(control(W)).
14 legal(oplayer, noop) :-
15       true(control(xplayer)).
16 ...
17 next(cell(M, N, x)) :-
18    does(xplayer, mark(M, N)),
19    true(cell(M, N, b)).
20 next(control(oplayer)) :-
21       true(control(xplayer)).
22 ...
23 row(M,X) :-
24    true(cell(M, 1, X)),
25    true(cell(M, 2, X)),
26    true(cell(M, 3, X)).
27 ...
28 line(X) :- row(M, X).
29 line(X) :- column(M, X).
30 line(X) :- diagonal(X).
31 ...
32 goal(xplayer, 100) :- line(x).
33 goal(xplayer, 0) :- line(o).
34 ...
35 terminal :- line(x).
```

Fig. 1. A partial GDL game description for the game Tic-Tac-Toe (reserved GDL keywords are marked in bold)

3. *If p and q occur in a cycle in Γ and G contains a clause*

$$p(s_1, \ldots, s_m) \Leftarrow b_1(\bar{t}_1) \wedge \ldots \wedge q(v_1, \ldots, v_k) \wedge \ldots \wedge b_n(\bar{t}_n)$$

then for every $i \in \{1, \ldots, k\}$,
  – v_i is variable-free, or
  – v_i is one of $s_1, \ldots, s_m$, or
  – v_i occurs in some $\bar{t}_j$ $(1 \leq j \leq n)$ such that b_j does not occur in a cycle with p in Γ.

4   Restrictions

As mentioned in [8] (Subsect. 3.2), there is no finite grounding of a GDL description in general. While the restrictions from Definition 1 ensure that reasoning about single states or state transitions is finite, the restrictions are not strong enough to ensure finiteness or decidability of reasoning about the game in general, such as, whether the game will terminate or is winnable for some player.

In fact, without further restrictions, GDL as defined in [11] or [18] is Turing complete [13]. Thus, some restrictions to the language are necessary in order to be able to ground a game description and only game descriptions that adhere to these restrictions can be grounded. The restriction used in [8] and termed *bounded GDL* in [13] is the following:

Definition 2. *Let G be a GDL specification. Let G' be G extended with the following three rules:*

1	$true\,(F)$	$:-\ init\,(F)$.
2	$true\,(F)$	$:-\ next\,(F)$.
3	$does\,(R,M)$	$:-\ legal\,(R,M)$.

G is in the bounded GDL *fragment of GDL descriptions, if, and only if, G' satisfies the recursion restriction.*

As discussed in [13], this restriction makes bounded GDL decidable and therefore truly less expressive than (unbounded) GDL. However, this is of little practical consequence as all of the game descriptions currently available in GDL belong to the bounded fragment.

5   Grounding

To obtain a ground version of a game description, we transform it into an answer set program P, ground P using very optimized grounder for answer set programs and extract the ground version of the game rules from the grounded answer set program.

Specifically, the program P that we create consists of

- the game description itself,
- a state generator,
- an action generator, and
- rules that encode all possible state terms and moves in the game.

The following definitions are based on [8], but simplified for our purpose.

Definition 3. *A* state generator *for a valid GDL specification G is an answer set program P^{gen} such that*

- *The only atoms in P^{gen} are of the form* true(f), *where $f \in \Sigma$, or auxiliary atoms that do not occur elsewhere; and*

– *for every reachable state S of G, P^{gen} has an answer set $\mathcal{A}$ such that for all $f \in \Sigma$: true(f) $\in \mathcal{A}$ iff $f \in S$.*

We use the following state generator

```
1    {true(F):base(F)}.
```

where, intuitively, *base(f)* encodes all possible terms f that might appear in a state of the game.

Definition 4. *Let $A(S)$ denote the set of all legal joint moves in S, that is,*

$$A(S) \overset{\text{def}}{=} \{A : R \mapsto \Sigma | l(r, A(r), S)\}$$

An action generator *for a valid GDL specification G is an answer set program P^{legal} such that*

– *The only atoms in P^{legal} are of the form* does(r, m), *where $r \in R$ and $m \in \Sigma$, or auxiliary atoms that do not occur elsewhere;*
– *for every reachable (non-terminal) state S of G and every joint move $A \in A(S)$, P^{legal} has an answer set $\mathcal{A}$ such that for all $r \in R$:* does($r, A(r)$) $\in \mathcal{A}$; *and*
– *for every reachable (terminal) state $S \in T$ of G, P^{legal} has an answer set.*

We use the following action generator

```
1    1={does(R, M):input(R, M)} :- role(R).
```

where, intuitively, *input(r, m)* encodes all possible moves m of role r in the game. Thus, our action generator does admit answer sets that might not be legal joint moves for a specific state. However, this is not a problem, since we are not interested in the answer sets, but only the grounded answer set program.

For several years, games in the international general game playing competition contain definitions of **base** and **input** predicates as used above. However, there is no formal definition of the semantics of those predicates in GDL and many older game descriptions do not have those predicates. We argue for the following definition:

Definition 5. *A game description G is said to have* well defined base and input definitions *if, and only if,*

– *for every reachable state S of G, for every $f \in S$, $G \vdash$ base(f); and*
– *for every reachable state S of G with $S \notin T$, role $r \in R$ and move $m \in \Sigma$, if $l(r, m, S)$ then $G \vdash$ input(r, m).*

Instead of putting the burden of writing well defined base and input definitions, we propose to generate them from the remaining rules. The idea for this is that the possible instances of *base(f)* should comprise all possible instances of *init(f)* and all possible instances *next(f)* for all possible state transitions.

Similarily, the possible instances of $input(r, m)$ must contain all instances of $legal(r, m)$ in any reachable non-terminal state.

The idea is to compute a *static* version P^{base} of the rules that define next states and legal moves of the players. Here, static means a relaxation of the rules that is independent of *true* and *does*, defined as follows.

Definition 6. *Let G be a GDL specification. We call a predicate p in G static iff $p \notin \{\text{init}, \text{true}, \text{next}, \text{legal}, \text{does}\}$ and p does neither depend on* true *nor* does *in the dependency graph of G.*

Furthermore, let p^{static} be a predicate symbol which represents a unique name for the static version of predicate p. By definition

$$\text{init}^{static} = \text{base}$$
$$\text{true}^{static} = \text{base}$$
$$\text{next}^{static} = \text{base}$$
$$\text{does}^{static} = \text{input}$$
$$\text{legal}^{static} = \text{input}$$
$$p^{static} = p \text{ if } p \text{ is static}$$

For each rule $p(\boldsymbol{X}) :- B \in G$ such that $p \in \{\text{init}, \text{next}, \text{legal}\}$ or either one of init, next, legal *depends on p in the dependency graph of G with positive edges, P^{base} contains the rule*

$$p^{static}(\boldsymbol{X}) :- B^{static}.$$

where B^{static} comprises the following literals:

$$\{q^{static}(\boldsymbol{Y}) : q(\boldsymbol{Y}) \in B\} \cup$$
$$\{\text{not } q(\boldsymbol{Y}) : \text{not } q(\boldsymbol{Y}) \in B \land q \text{ is static }\}$$

As an example, the following rules form P^{base} as generated for the Tic-Tac-Toe game (Fig. 1):

```
1  base(cell(1, 1, b)).
2  ...
3  base(cell(3, 3, b)).
4  base(control(xplayer)).
5  base(cell(M, N, x))  :-
6      input(xplayer, mark(M, N)),
7      base(cell(M, N, b)).
8  base(cell(M, N, o))  :-
9      input(oplayer, mark(M, N)),
10     base(cell(M, N, b)).
11 base(cell(M, N, C))  :-
12     base(cell(M, N, C)), C!=b.
13 base(control(xplayer))  :-
14     base(control(oplayer)).
15 base(control(oplayer))  :-
16     base(control(xplayer)).
```

```
17 base(cell(M, N, b)) :-
18     base(cell(M, N, b)),
19     input(R, mark(X, Y)), M!=X.
20 base(cell(M, N, b)) :-
21     base(cell(M, N, b)),
22     input(R, mark(X, Y)), N!=Y.
23
24 input(W, mark(X, Y)) :-
25     base(control(W)), base(cell(X, Y, b)).
26 input(xplayer, noop) :-
27     base(control(oplayer)).
28 input(oplayer, noop) :-
29     base(control(xplayer)).
```

The answer set program P that we generate from a game description G is defined as $P = G \cup P^{base} \cup$

```
1    {true(F):base(F)}.
2    1={does(R, M):input(R, M)} :- role(R).
```

As can easily be seen, our definition of P^{base} (and thus P) fulfills the restrictions for a valid GDL description (Definition 1) if the original game description G belongs to the bounded fragment of GDL (Definition 2). The reason is that in P^{base} we introduce recursions involving $true^{static}$ and $next^{static}$ (which are both base) and therefor also between $legal^{static}$ and $does^{static}$ (both input). Thus, all bounded GDL programs can be grounded using this method in principle. GDL descriptions not fulfilling the restrictions for bounded GDL, can lead to an infinite ground representation.

That said, grounding bounded GDL can still lead to an exponential blowup in the size of the representation which can make grounding infeasible. Especially games containing rules with many variables suffer from this problem.

Optimizations. Before grounding the answer set program P, we apply optimizations to it similar to the ones described in [8] (Subsect. 6.2). That is, we try to reduce the resulting grounding by removing existential variables and removing unnecessary rules, as illustrated in the following paragraphs.

As an example, consider the rule p(X,Z) :- q(X,Y), r(Y), s(Z).. The variable Y in the body is existentially quantified (does not appear in the head). We replace this rule by

```
1    p(X,Z) :- qr(X), s(Z).
2    qr(X)  :- q(X,Y), r(Y).
```

where qr is a new predicate symbol and obtain two rules with two variables each instead of one rule with three variables. This reduces the number of ground rules that are generated (unless the domains of the variables are singletons).

Some rules in the game descriptions are unnecessary and can be removed. For example, Tic-Tac-Toe (see Fig. 1) contains the rules for line(X). In those rules X can be replaced with any of $\{x, o, b\}$, however only line(x) and line(o) appear in the body of another rule. Thus, the ground rules that would be generated for `line(b)` are irrelevant as are the ground instances of `row`, `column` and `diagonal` where X is replaced with b. We prevent these unnecessary rules from being generated in the first place, by instantiating the X in the rules for `line(X)` with x and o and handing these partially instantiated rules to the grounder.

6   Experiments

We ran experiments on 231 games from the GGP server [15]. For each game we ran our grounder and recorded

- the time it took to generate the answer set program P;
- the total runtime for grounding (including the time for generating P);
- the size of the resulting ground description in terms of number of resulting clauses;
- the number of components of a propositional network created from the those clauses without optimizations.

For comparison, we used the GGP-Base framework [20] to generate a propositional network (propnet) using the OptimizingPropNetFactory class. Generating a propnet includes grounding the game description as a first step and we measured only the time for this step without the remaining time that is spent on optimizing the propositional network. However, these two are somewhat intertwined such that a complete separation is not possible. GGP-Base makes use of `base` and `input` predicates. Since most games on the GGP server do not contain `base` and `input` definitions, we added the base and input definitions that were generated by our own grounder to the game rules that were given as input to the propnet generation. For GGP-Base we recorded the runtime and the number of resulting components, where each component represents a conjunction, disjunction, negation or proposition in the (grounded) rules. Thus, this number is roughly comparable to the number of clauses (including facts) in the grounded game description.

The ASP-based grounder can ground 226 of the 231 tested games within the time limit of 1 h and memory limit of 4 GB. The median runtime was 1.4 s (average 4.5 s), which includes the time for starting an external process for the grounder and reading the resulting grounded game description. For comparison, the GGP-Base grounder can ground 218 of the tested games with a median runtime of 2.4 s (average 5.9 s). Since no external process needs to be started, this runtime does not include any process communication overhead. There was no game that could be grounded using GGP-Base but not using ASP. Most of the games that could be handled by the ASP-based grounder but not by GGP-Base feature heavy use of recursive rules. Neither system could ground laik-Lee_hex, merrills, mummymaze1p, ruledepthquadratic or small_dominion. All of those games feature recursive rules, except for mummymaze1p, which could be grounded by the ASP system resulting in about 5 million clauses, but processing the ground clauses and generating the propnet took too much time. The games

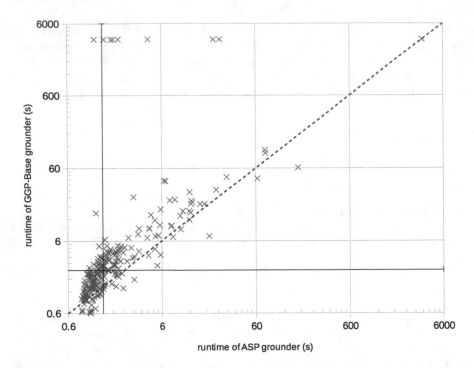

Fig. 2. Runtime for grounding game descriptions with GGP-Base vs. our ASP-based grounder. Points above the diagonal denote games where the GGP-Base grounder takes longer than the ASP-based grounder. The runtime limit was set to 1 hour. Thus, points at x = 3600 s denote games where the ASP grounder failed to ground the game within the time limit (and conversely for the GGP-Base grounder). The horizontal and vertical lines show the median runtime of GGP-Base and ASP-based grounder, respectively.

farmers, god, Goldrush, kalaha_2009, quad_5x5, SC_TestOnly, sudoku_simple and uf20-01.cnf.SAT could be grounded by the ASP-based grounder, while the GGP-Base grounder exceeds the run-time limit. Of those games, only farmers and kalaha_2009 can be considered complex games taking 25.4 s and 21.7 s to ground respectively and resulting in more than 100000 components. The other games could all be grounded by the ASP-based grounder in under 5 s resulting in no more than 13000 components.

Figure 2 shows the runtimes of the GGP-Base vs. ASP-based grounder for all 231 games on a logarithmic scale. We can see that most games can be grounded in less than 1 min with both grounders. In 173 cases the ASP-based grounder is faster than the GGP-Base grounder (some of them within the margin of error). On average, the ASP-based grounder is 20.4% faster than the GGP-Base grounder.

We compared the size of the grounded description with both systems in Fig. 3. As can be seen, there is little difference between both systems, but the GGP-Base system creates significantly larger groundings in few selected games (smallest,

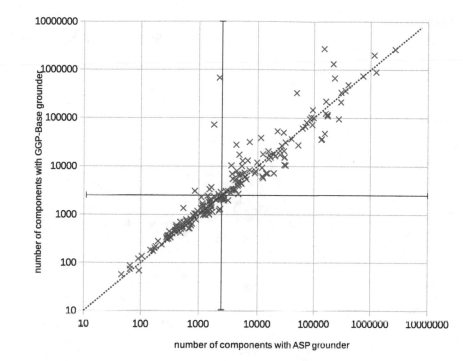

Fig. 3. Number of components of a propositional network created from the grounding resulting from GGP-Base vs. our ASP-based grounder. Points above the diagonal denote games where the GGP-Base grounder creates larger propnets. The horizontal and vertical lines show the median numbers of components for GGP-Base and ASP-based grounder, respectively.

logistics, mastermind, crossers3, othello-comp2007, othellosuicide, racer, racer4, battlebrushes). The number of clauses of the grounded game descriptions range from 26 (from troublemaker01) to 2038583 (for battlesnakes1509) with a median of 2444. The number of components in the generated propositional networks is similar (between 46 and 2715284). The median number of generated components is 2455 for the ASP-based grounder vs. 2518 for the GGP-Base grounder.

In Fig. 4, we plotted the size of the ground representation (propositional network) compared to the size of the original GDL rules for each of the games. We used the smaller of the two groundings for each game and measured the size of the rules by taking the sum of the number of literals of all rules. As can be seen in the graph, the propositional network is generally some orders of magnitude larger than the GDL rules, except for some edge cases where the rules are essentially already grounded. However, we cannot see a general trend indicating an exponential blowup in size. That is, although this blowup is theoretically possible, it seems to happen rarely in the games we looked at. In fact, the best fit to the data (excluding the abnormal cases) seems to be a power law which puts the propnet size at slightly more than a quadratic function of the size of the GDL rules.

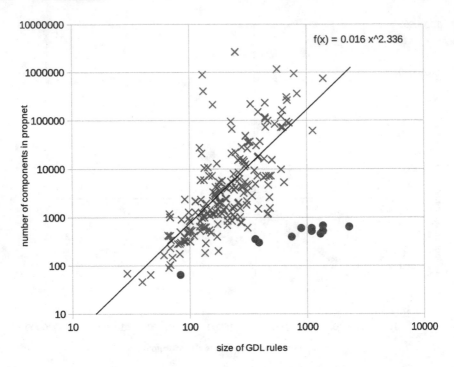

Fig. 4. Number of components of the propositional network compared to the size of the original GDL description (without base and input rules) measured in number of literals for all 226 games that could be grounded. The dots denote games that are abnormal in that the GDL rules are larger than the propnet. All of these games turned out to be either test case games made to test certain aspects of GDL reasoners (as opposed to general game players) or games that are essentially already ground. The line shows the best matching regression.

7   Conclusion

Grounding game descriptions using a state-of-the art answer set programming system is a viable alternative to the GDL specific approach implemented in the GGP-Base framework. The system we presented is able to handle more games and is typically faster despite the overhead of transforming GDL into a different format and starting and communicating with a separate process. Furthermore, our grounding of a game description is well-founded theoretically by the transformation into answer set programs. This allows to optimize the descriptions further without changing their semantics. In the future, we plan to look into further optimizations of the grounding to allow grounding of more complex game descriptions. Additionally, these optimizations will likely reduce the size of the grounded descriptions which generally leads to faster reasoning with the grounded game descriptions, for example, in the form of propositional networks. However, even with those optimization there will likely be games where the potential exponential blowup will prevent grounding from being feasible.

In those cases it is necessary to fall back on reasoners that do not require a propositional representation (e.g., Prolog).

Acknowledgments. This work was supported by the Icelandic Centre for Research (RANNIS).

References

1. Apt, K., Blair, H.A., Walker, A.: Towards a theory of declarative knowledge. In: Minker, J. (ed.) Foundations of Deductive Databases and Logic Programming, chap. 2, pp. 89–148. Morgan Kaufmann (1987)
2. Baral, C.: Knowledge Representation, Reasoning, and Declarative Problem Solving. Cambridge University Press, New York (2003)
3. Cerexhe, T., Rajaratnam, D., Saffidine, A., Thielscher, M.: A systematic solution to the (de-)composition problem in general game playing. In: Proceedings of the European Conference on Artificial Intelligence (ECAI), pp. 195–200 (2014)
4. Cox, E., Schkufza, E., Madsen, R., Genesereth, M.: Factoring general games using propositional automata. In: Proceedings of the IJCAI-09 Workshop on General Game Playing (GIGA 2009), pp. 13–20 (2009)
5. Edelkamp, S., Kissmann, P.: On the complexity of BDDs for state space search: a case study in Connect Four. In: Proceedings of the AAAI Conference on Artificial Intelligence, pp. 18–23. AAAI Press (2011)
6. Gebser, M., Kaminski, R., Kaufmann, B., Ostrowski, M., Schaub, T., Schneider, M.: Potassco: the potsdam answer set solving collection. AI Commun. **24**(2), 107–124 (2011)
7. Van Gelder, A.: The alternating fixpoint of logic programs with negation. In: Proceedings of the 8th Symposium on Principles of Database Systems, pp. 1–10. ACM SIGACT-SIGMOD (1989)
8. Haufe, S., Schiffel, S., Thielscher, M.: Automated verification of state sequence invariants in general game playing. Artif. Intell. **187–188**, 1–30 (2012)
9. Kissmann, P., Edelkamp, S.: Instantiating general games using prolog or dependency graphs. In: German Conference on Artificial Intelligence, pp. 255–262 (2010)
10. Lloyd, J., Topor, R.: A basis for deductive database systems II. J. Logic Program. **3**(1), 55–67 (1986)
11. Love, N., Hinrichs, T., Haley, D., Schkufza, E., Genesereth, M.: General game playing: Game description language specification. Technical report, Stanford University (2008). http://games.stanford.edu/language/spec/gdl_spec_2008_03.pdf
12. Michulke, D., Schiffel, S.: Admissible distance heuristics for general games. In: Filipe, J., Fred, A. (eds.) ICAART 2012. CCIS, vol. 358, pp. 188–203. Springer, Heidelberg (2013). doi:10.1007/978-3-642-36907-0_13
13. Saffidine, A.: The game description language is Turing complete. IEEE Trans. Comput. Intell. AI Games **6**(4), 320–324 (2014)
14. Schiffel, S.: Symmetry detection in general game playing. In: Proceedings of the AAAI Conference on Artificial Intelligence, pp. 980–985. AAAI Press (2010)
15. Schiffel, S.: GGPServer (2016). http://ggpserver.general-game-playing.de/
16. Schiffel, S., Björnsson, Y.: Efficiency of GDL reasoners. IEEE Trans. Comput. Intell. AI Games **6**(4), 343–354 (2014)
17. Schiffel, S., Thielscher, M.: Automated theorem proving for general game playing. In: Proceedings of IJCAI 2009, pp. 911–916 (2009)

18. Schiffel, S., Thielscher, M.: A multiagent semantics for the game description language. In: Filipe, J., Fred, A., Sharp, B. (eds.) ICAART 2009. CCIS, vol. 67, pp. 44–55. Springer, Heidelberg (2010). doi:10.1007/978-3-642-11819-7_4
19. Schkufza, E., Love, N., Genesereth, M.: Propositional automata and cell automata: representational frameworks for discrete dynamic systems. In: Wobcke, W., Zhang, M. (eds.) AI 2008. LNCS (LNAI), vol. 5360, pp. 56–66. Springer, Heidelberg (2008). doi:10.1007/978-3-540-89378-3_6
20. Schreiber, S., Landau, A.: The general game playing base package (2016). https://github.com/ggp-org/ggp-base

A General Approach of Game Description Decomposition for General Game Playing

Aline Hufschmitt(✉), Jean-Noël Vittaut, and Jean Méhat

LIASD - University of Paris 8, Saint-Denis, France
{alinehuf,jnv,jm}@ai.univ-paris8.fr
http://www.ai.univ-paris8.fr

Abstract. We present a general approach for the decomposition of games described in the *Game Description Language* (GDL). In the field of *General Game Playing*, the exploration of games described in GDL can be significantly sped up by the decomposition of the problem in sub-problems analyzed separately. Our program can decompose game descriptions with any number of players while addressing the problem of joint moves. This approach is used to identify perfectly separable sub-games but can also decompose serial games composed of two subgames and games with compound moves while avoiding, unlike previous works, to rely on syntactic elements that can be eliminated by simply rewriting the GDL rules. We tested our program on 40 games, compound or not, and we can decompose 32 of them successfully in less than 5 s.

1 Introduction

Despite incentives from Genesereth and Björnsson [3] to encourage the development of GGP players able to discern structure of compound games and therefore to dramatically decrease search cost, not much research exists in this area.

Cox et al. [2] prove conditions under which a global game represents multiple, simultaneous independent sub-games, but the practical implementation of a GGP player using decomposition presents two major issues: the first is to detect and decompose a compound game, the second is to combine local subgame solutions into a global one.

Cerexhe et al. [1] provide a systematic approach for single player games to solve this second difficulty which they refer to as the *composition problem*. However, identifying and decomposing games is not within the scope of their paper.

Günther et al. [5,6] propose a decomposition approach for single player games by building a dependency graph between fluents and actions: the connected parts of the graph represent the different subgames. Potential preconditions, positive and negative effects between fluents and actions are used to build this dependency graph while action-independent fluents are isolated in a separate subgame to prevent them from blocking the decomposition.

Zhao et al. [10,11] propose a similar approach for multiplayer games using partially instantiated fluent and action terms. Serial games and games with compound actions are handled separately.

© Springer International Publishing AG 2017
T. Cazenave et al. (Eds.): CGW 2016/GIGA 2016, CCIS 705, pp. 165–177, 2017.
DOI: 10.1007/978-3-319-57969-6_12

These approaches present different shortcoming we will details below such as a heavy reliance on certain syntactic structures in game descriptions.

We propose a more general approach to decompose games with any number of players while addressing the problem of joint moves, compound moves and serial games without relying on syntactic elements that can be eliminated by simply rewriting the GDL rules. The result of our decomposition can be used to solve the game by an approach like the one of Cerexhe et al. [1]; it is a non-trivial problem outside the scope of this paper.

We begin (Sect. 2) with a brief introduction of the *Game Description Language* and the different types of compound games that can be found on the different online servers and that our approach can decompose. Then we present the different aspects of our method to handle these different types of games (Sect. 3). We present results on 40 games, compound or not (Sect. 4). Finally, we conclude and present future work (Sect. 5).

2   Preliminaries

We present here some details about the *Game Description Language* and the different types of compound games that our approach can decompose.

2.1   The Game Description Language

We assume familiarity of the reader with the General Game Playing [4] as well as with the Game Description Language (GDL) [7]. A GDL game description takes the form of a set of assertions and of logical rules which conclusion describes: the transition to the next position (*next* predicate); the legality of actions (*legal*); the game termination (*terminal*); and the score (*goal*). The rules are expressed in terms of actions (*does*) and fluents (*true*) describing the game state.

Rule premises can also include *auxiliary predicates*, specific to the game description itself, which truth is defined by rules also using *true* and *does* premises. In the rest of this article, we will refer to *auxiliary predicates*, exclusively defined in terms of fluents (*true*) (*does* never appear in their premises), which have an important role in our decomposition approach (Sects. 3.3, 3.5).

2.2   Types of Compound Games

Among games available on the different *General Game Playing* servers (http://games.ggp.org) different types of compound games can be identified. The types we distinguish represent specific issues for the decomposition and are not directly related to the formal classification proposed by (Cerexhe et al. [1]).

For example, **Parallel games** like *Dual Connect 4* or *Double Tictactoe Dengji* are composed of two subgames played in parallel that can be *synchronous* or *asynchronous*, but this difference has no influence on the decomposition approach to use. Decomposing these games does not present any particular difficulty.

However, in some synchronous parallel games like *Asteroids Parallel* each player's action is a **compound moves** corresponding to two simultaneous actions played in each subgames. These create a strong connection between subgames and represent a specific difficulty for decomposition.

Serial Games like *Blocker Serial* are composed of two *sequential* subgames i.e. the second starts when the first is completed. As the two games are linked together, identifying the boundary between them is a specific issue for decomposition.

Multiple Games like *Multiple Buttons And Lights* are composed of several subgames, only one of them being involved in the score calculation or the game termination. The other subgames only increase the size of the game tree to explore. Identifying those *useless* subgames allows to avoid unnecessary calculations. Note that in the game *Incredible*, *contemplate* actions are detected as *noop* actions by our decomposition program and does not constitute a *useless* subgame.

Games using a Stepper to ensure finite games like *Eight Puzzle* may be considered as compound games (*synchronous*). In these games, different descriptions of a position can vary only by the value of the *stepper* (step counter). To allow a programmed player to exploit these near-perfect transpositions, it is necessary to operate a game decomposition to separate the stepper from the game itself. This stepper is then an *action independent subgame*.

Some Impartial Games, like *Nim* starting with several piles of objects, may also be considered as compounds games (*asynchronous*) as they can be decomposed in several subgames, one for each pile, each of them being an impartial game [10]. Identifying that these subgames are impartial, subsequently allows to use known techniques for the resolution of the global game.

3   Method

Our approach is based on Günther's idea [5] and consists in using a dependency graph between actions and fluents, and then to identify the connected parts of the graph representing the subgames. As nothing in the GDL specification prohibits the use of completely instantiated rules or prevents that fluents or actions be reduced to simple atoms, we identify relations between totally instantiated fluents f and actions a and rely neither on their predicates names nor their arguments.

For the analysis of these relations, we use the following definitions:

Definition 1. *Let F be the set of all the instantiated fluents f appearing in $true(f)$ or $\neg true(f)$.*

Definition 2. *R being the set of all the roles r and O the set of all options o of these roles, let $A \subset R \times O$ be the set of all the instantiated player actions $a = (r, o)$.*
O_r is the set of all the possible options of role r.

Definition 3. *Let C be the set of all the possible conjunctions of atoms of the form $true(f)$, $\neg true(f)$, $does(r, o)$ or $\neg does(r, o)$.*

3.1   Grounding and Creation of a Logic Circuit

To instantiate completely the rules (grounding), we carry out a fast instantiation using Prolog with tabling [9] and use these instantiated rules to build a logic *circuit* similar to a *propnet* [8]. Conclusions of *legal*, *next*, *goal* or *teminal* rules are the outputs of the circuit and only depends on fluents (*true*) and actions (*does*) at the inputs.

It is possible, according to the GDL specifications, to produce a description with fully developed rules using no auxiliary predicate at all. However, these predicates, like *column1*, *diagonal2* or *game1over* in *Tictactoe*, may be necessary for some specific stages of our process of decomposition (Sects. 3.3, 3.5). To ensure that these auxiliary predicates will be available even when not specified in the GDL description, we proceed to a factorization of the conjunctions, disjunctions and use De Morgan's laws to reduce the number of negations in the circuit. As a perfect factorization is an NP-hard problem, our program uses a greedy approach where the first common factor is used. Factorization and application of De Morgan's laws are iterated until the circuit reaches a minimum size.

We identify the needed auxiliary predicates as these are represented by internal logic gates of the circuit, depending only on input fluents and representing important expressions in the logic of the game i.e. these expressions are used several times, several logic gates use their outputs.

After the factorization, the GDL description is a set of formulas under disjunctive normal form of which atoms are fluents, actions, and auxiliary predicates. In the following we say that these formulas are under *DNF* form.

Other stages of the decomposition process need a description of the game under canonical form. By recursively replacing auxiliary predicates by their expression we obtain a new set of formulas in disjunctive normal form describing the same game where all the auxiliary predicates have been eliminated. In the following we say that these formulas are under *DNFD* form.

3.2   Building a Dependency Graph

To build our dependency graph and to identify the different subgames, we start with a set of vertices which are the fully instantiated actions and fluents. We then identify different relations between these fluents and actions that we define below. For each of these relations we add an edge between the involved actions and fluents vertices. These relations correspond to preconditions or effects of the actions.

Unfortunately, GDL does not explicitly describe action effects unlike STRIPS or PDDL languages used for planning domains. A fluent being *false* by default, an action present in a *next* rule can have an effect or not. For example, let us consider the legal actions $does(r, a)$, $does(r, b)$ and $does(r, c)$, in the rule $next(f):- \neg true(f) \wedge (does(r, a) \vee does(r, b))$. a and b have an effect if the rule means *"The cell will contain a pawn if r does one of the 2 actions moving a pawn in it"* and c has an effect if it means *"the boat will sink if r does anything else than action c (bailing)"*. A similar example can be found for any *next* rule with an action (in a negation or not) and regardless of the value of the fluent f and its presence or not in the rule premises.

It is thus possible to produce GDL descriptions in which the actions present in a *next* rule body belong to another subgame than the fluent in the rule head. We can only address this using heuristics similar to those of Günther [6].

They propose to consider that an action a has a negative effect on a fluent f if this action does not keep the fluent true i.e. if $next(f)$ does not contain $true(f) \wedge does(a)$ in its premises. However in a game like Double Tictactoe, there is no rule like this to indicates that actions of a subgame do not change the value of the other subgame fluents. Consequently, fluents of a subgame can be considered as negative effects of the second subgame actions and the decomposition fails.

In our approach we use slightly different heuristics which work well for existing composed games to find potential effects of actions:

Definition 4. *The fluent f is a **potential negative effect** of the action $a = (r, o)$ if $next(f)$ under DNFD has a clause where $\neg does(r, o)$ appears.*

*The fluent f is a **potential positive effect** of the action $a = (r, o)$ if $next(f)$ under DNFD has a clause containing the $does(r, o)$ literal and not containing the $true(f)$ literal.*

In case of joint moves from several players, it is necessary to identify if the action of each player is responsible of the observed effect on the rule conclusion to avoid linking unrelated action with the conclusion.

To solve this problem Zhao et al. [11] propose to compare the arguments used in a *next* rule head with the ones used in the moves (*does*). For example, in the following rule from *Blocker Serial*, we can see that the action from *crosser* is the only one that is likely to affect the conclusion:

$$next(cell2(\mathbf{XC}, \mathbf{YC}, crosser)) :- distinctcell(\mathbf{XC}, \mathbf{YC}, XB, YB)$$
$$\wedge\ does(crosser, mark2(\mathbf{XC}, \mathbf{YC})) \wedge\ does(blocker, mark2(XB, YB)).$$

However, GDL specification allows to use completely instantiated rules and simple atoms to represent fluents and moves. For example, we can replace the previous rule by some instantiated rules:

$$next(f) :- does(crosser, o1) \wedge does(blocker, o2).$$
$$next(f) :- does(crosser, o1) \wedge does(blocker, o3).$$

$$\cdots$$

With fluents like f and moves like $does(r, o)$, their approach is no longer able to deal with joint moves.

To identify which action has an effect without relying on syntactic elements, we compare, for each player, the different actions used in conjunction with the same fluents and actions of other players in the clauses of each *next* rule.

Suppose that $next(f) \leftarrow C_f$ is in DNFD. Let us consider a specific option o' for player r'. We consider the set $E(o')$ of the different options of the role r when r' choose the o' option:

$$E(o') = \{ \ o \in O_r \ \mid \ \exists c \in C_f, \exists b \in C,$$
$$c = does(r,o) \wedge does(r',o') \wedge b \ \}$$

We define $E(o)$ the same way by exchanging the role of (r,o) with (r',o').

If all the options of the r are present in conjunction with the same action of r': these options have probably no effect i.e. the result is the same regardless of the option chosen. On the contrary, if a single option of r is present, it is probably responsible for the observed effect. We then use the following heuristics:

Definition 5. *The action* $a = (r,o) \in A$ *is **potentially responsible for an effect** on f if:*

- $card(E(o')) = 1$, *or*
- $E(o') \subsetneq O_r$ *and* $card(E(o)) \neq 1$

For example, in the game *BlockerSerial*, the term $next(cell1(2,3,crosser))$ is *true* if *blocker* choose any option but $mark1(2,3)$ and *crosser* choose the $mark1(2,3)$ option. All the options of *blocker* are not represented but, as *crosser* has a single possible option, its action is considered responsible for the effect while actions of *blocker* are not linked to the $cell1(2,3,crosser)$ fluent.

Even if this approach sometimes put aside actions related to the conclusion, we did not observe any over-decomposition. At least one of the actions is indeed related to the conclusion and edges between fluents and actions added in the dependency graph to represent preconditions relations are redundant with those added for effect relations.

Therefore a fluent is a **potential effect** of an action if this action has a *potential positive or negative effect* on this fluent and if this action is *potentially responsible for this effect* in presence of joint moves. From the potential effect of actions we can deduce fluents that are action-independent, such as *step* or *control* fluents, and actions that are fluent-independent such as *noop* actions:

Definition 6. *A fluent f is **action-independent** if it is not the potential effect of any action a. An action a is **fluent-independent** if no fluent f is the potential effect of this action.*

Then we can identify fluents that are potential preconditions of an action in the same subgame and create a link in the graph between them:

Definition 7. *The fluent f is a **potential precondition in the same subgame** of the action* $a = (r,o)$ *if:*

- *a is not fluent-independent, and*
- *f is not action-independent, and*
- *one of the two following conditions holds:*
  - *$legal(r,o)$ under DNFD has a clause where $true(f)$ or $\neg true(f)$ appears, or*
  - *it exist f' which is a potential effect of a, such that $next(f')$ under DNFD has a clause containing $does(r,o) \wedge true(f)$ or $does(r,o) \wedge \neg true(f)$.*

An action-independent fluent can be present in the premises of all *legal* rules, it is then a precondition of all actions but belongs to another subgame which is action-independent.

3.3   Subgoal-Predicates to Fix Over-Decomposition

Edges between actions and fluent vertices corresponding to preconditions or effects of these actions may not be sufficient to connect all the elements of a subgame. For instance, in a subgame like *Tictactoe*, an action has an effect on a cell and the state of this cell is a precondition to this action. However, no link exists through actions between fluents describing different cells.

In the game *Double Tictactoe* given as an example by Zhao et al. [11] the *auxiliary predicates line1/1* or *line2/1* are present in the premises of some *legal* rules. All the fluents in the premises of these predicates are then preconditions of the corresponding actions and create a link between the cells of each subgame. However, in games like *Tictactoe Parallel*, *Connect4* or *Rainbow* no such predicate is present in the *legal* rules and an over-decomposition occurs.

The logic link between elements of a subgame is in the goal to reach and this goal is usually a condition for the termination of the global game. We need to distinguish an auxiliary predicate corresponding to a subgoal in one subgame from one corresponding to different subgoals from different subgames because the second one can prevent the decomposition. To address this problem of over-decomposition we use the following heuristic to identify potential subgoal-predicates corresponding to only one subgame:

Definition 8. *Let g be the maximum possible score of r. An auxiliary predicate b is a **potential subgoal-predicate** if:*

- *terminal depends on the logical value of b, and*
- *goal(r, g) under DNF has a clause where b appears.*

or

- *All the roles play in different subgames, and*
- *goal(r, g) under DNF has a clause where b appears, and for all roles $r' \neq r$, goal(r', g') under DNF has no clause where b appears.*

In games like *Dual Rainbow* or *Dual Hamilton*, subgoal-predicates appear only in the premises of *goal* rules. Since these games are composed of single player subgames, an auxiliary predicate present in the *goal* rule of a single player involves only this player and therefore only one subgame.

The first part of the definition holds in the games where the victory in one of the subgames terminates the game as it is generally the case in compound games. Otherwise, the subgames may be connected by the use of a misidentified subgoal-predicate.

Once a subgoal-predicate is identified, we add edges in our dependency graph between fluents that appear in a same clause in its formulas under DNFD.

3.4   Compound Moves and Meta-Action Sets

A compound move is composed of two or more actions related to different subgames. For example, the compound move *legal(ship,do(clockcounter))* in the game *Asteroid Parallel* corresponds to a *clock*wise move in a first subgame and a *counter*clockwise move in a second subgame. Such an action creates a link between the different subgames and can interfere with the decomposition process.

To detect compound moves, Zhao et al. [11] use the same app-roach as that applied to the problem of joints move. For example, in the following rule from *Tictactoe Parallel* we can see that only the first two arguments of the action have an effect on the rule conclusion: $next(cell1(\mathbf{X1}, \mathbf{Y1}, o))$:− $does(oplayer, mark(\mathbf{X1}, \mathbf{Y1}, X2, Y2))$. Once again, the rule has just to be rewritten to defeat detection: $next(f)$:− $does(oplayer, o)$.

In games with compound moves, the set of all actions is a combination of the sets of all actions of each subgame. Then in a game composed of two subgames, for each action in the first subgame, there is N compound moves corresponding to this action combined to the N possible actions in the second subgame. To identify the different parts of compound moves, we distribute actions into meta-action sets. An action can belong to one or several meta-action sets which depend only on a role r, a fluent $f \in F$ and two clauses $c \in C$ and $c' \in C$.

Definition 9. *An action* $a = (r, o)$ *belongs to the meta-action set* $P(r, f, c, c')$ *if:*

- *f is a potential effect of a, and*
- *$next(f)$ under DNFD has a clause $(does(r, o) \wedge c)$, and*
- *if c' is empty, $legal(r, o)$ must always be true, or if c' is not empty, it contains only action-dependent literals and appears in at least one clause of $legal(r, o)$ under DNFD.*

Therefore a meta-action set is a group of actions with an identical effect on a fluent of a particular subgame, the same preconditions in the corresponding *next* rule and at least one precondition in common in their *legal* rules.

For example, in the game *Blocks World Parallel* we can find the meta-action set $\{does(robot, do(\mathbf{stack}stack, \mathbf{a}, \mathbf{b}, *, *)), does(robot, do(\mathbf{stack}unstack, \mathbf{a}, \mathbf{b}, *, *))\}^1$ corresponding to the action $stack(a, b)$ in the first subgame. These actions have an effect in common on $true(on1(a, b))$, same preconditions $\{true(table1(a)), true(clear1(b)), true(clear1(a))\}$ in the $next(on1(a, b))$ clauses and are always legal.

In a game with compound actions, each action is placed in M meta-action sets corresponding to M effects. If a game contains no compound action but some actions with an identical effect in the same situation, these actions are grouped in the same meta-action set. And finally, if all actions in a game have a different effect, each one constitutes a meta-action singleton. The use of meta-action sets is then compatible with all games.

In our dependency graph, we then encapsulate all actions into meta-action sets to avoid compound actions from connecting different subgames. The links

---

[1] The * represents different possible values, the whole meta-action set contains 12 compound moves.

between actions and fluents are replaced by links between action sets and fluents i.e. in the dependency graph, edges are added between a meta-action set and its effect f and preconditions $f' \in c \cup c'$.

3.5   Serial Games

In serial games an auxiliary predicate describing the terminal situation of the first subgame determines the legality of all actions of the second subgame. Consequently, it creates links between first subgame fluents and second subgame actions. We must detect it and avoid these links to separate both subgames.

Zhao [10] uses a separate special detection: the desired auxiliary predicate must be *false* to authorize the first subgame actions and *true* to authorize the second ones, like *game1over* in *Tictactoe Serial*:

$$legal(PLAYER, mark1(X,Y)) :- \neg game1over \wedge \dots .$$
$$legal(PLAYER, mark2(X,Y)) :- game1over \wedge \dots .$$

with *game1over* depending on $line1(x) \vee line1(o) \vee \neg open1$. However, someone can defeat this approach by simply rewriting the first subgame *legal* rules with a different precondition: $legal(PLAYER, mark1(X,Y)) :- ongoing1 \wedge \dots$. with *ongoing1* depending on $\neg line1(x) \wedge \neg line1(o) \wedge open1$.

To generalize the approach of Zhao [10], we consider that a *pivot* between two serial subgames is composed of two auxiliary predicates that can be the negation of each other or two completely different predicates. We use our circuit representing the game to test the influence of each auxiliary predicate detected during the circuit creation on the actions legality and look for a couple of predicates that parts the fluent-dependent actions in two groups.

If such a couple of auxiliary predicates is found, then it is a *pivot* and the latter predicates are directly used as action preconditions instead of the fluents included in them. In our dependency graph, fluents of the first subgame are then encapsulated in these auxiliary predicates to ensure that they will not connect the different subgames with direct links to actions (meta-action sets) of the second subgame. This approach works for existing games that are limited to two serial subgames.

Unfortunately, we cannot generalize this approach and identify a *pivot* in case of more than two serial subgames without risking an over-decomposition of games with movable parts. In a *pivot*, each auxiliary predicate is necessary to allow the legality of some actions and may prevent the legality of other actions. If a third subgame is present, its actions are not affected by both auxiliary predicates. In a game with movable pawns, an auxiliary predicate may be used to describe the state of a cell; this predicate may allow the legality of some moves from this cell, prevent some moves to this cell and does not concern other moves of the game, consequently it may be confused with a part of a *pivot*. Therefore, if we try to identify *pivots* for more than two serial subgames with a generalization of this approach, a game with movable pawn may be over-decomposed, each cell being a small serial subgame leading to the next ones.

3.6 Multiple Games and Useless Subgames

Some subgames are involved in the calculation of the score or can cause the end of the game when some position is reached. A subgame may also be played to allow another subgame to start in the case of serial subgames.

Definition 10. *Let V_S be the set of vertices of a connected part of the dependency graph representing a subgame S. S is considered **useful** if:*

- *S is played before another subgame in a serial game and is necessary to start it, or*
- *it exists $f \in F \cap V_S$ such that terminal depends on the logical value of $true(f)$, or*
- *it exists $f \in F \cap V_S$ such that $goal(r, g)$ depends on the logical value of $true(f)$.*

In multiple games, all the subgames that are not identified as *useful* can be ignored and remain unexplored. However, a *useless* action (*noop*) can be sometime strategically useful to avoid a *zugzwang* in another subgame. Actions of these subgames can then be flagged as *noop* actions, be considered equivalently *useless*, and only one of them need to be explored (if legal) for each position of the game.

4   Experiments

We evaluated our decomposition program on a panel of 40 descriptions of games, compound or not, from the servers of Dresden, Stanford and Tiltyard. We took all the available compound games except for the redundant ones. We added the original version of games commonly used as subgames and a representative panel of games with different characteristics (movable parts, steppers, asymmetry, impartiality) and complexity. The experiments were run on one core of an Intel Core i7 2,7 GHz with 8Go of 1600 MHz DDR3.

For each game, we measured the mean time necessary for each stage of the decomposition on a set of 100 decomposition tests. To limit the duration of the experiments, a decomposition test was aborted after 60 min. The longest stages of the decomposition are grounding the rules, factorizing the circuit and calculating completely developed disjunctive normal forms (DNFD). The column 5 of Table 1 indicates the total time needed to decompose each game and shows that the DNFD calculation can be very time consuming.

We try to compute DNF without developing the auxiliary predicates identified during the circuit construction. As we can see it in column 6, the time saved is really significant and allows the successful decomposition of 32 games among 40 in less than 5 s. The major part of the total time necessary for the decomposition using DNF corresponds to the rules grounding and circuit factorization.

Unfortunately, the use of partially developed DNF presents a shortcoming: if a rule containing variables is already instantiated in the original GDL description of a game and if some of these instances only are expressed in terms of auxiliary predicates, actions may occur in conjunction with different but equivalent premises: a group of fluents or an equivalent auxiliary predicate. The factorization of the circuit should restore auxiliary predicates in all rules instances but as we use a greedy approach (Sect. 3.1), it is not guaranteed. Therefore, meta-action sets detection may be hindered. Nevertheless, this case is sufficiently specific to successfully use the auxiliary predicates in DNF, in most cases.

For *Hex* and *Blocker Parallel*, the time required to compute the grounded rules, the factorization and the DNFs still remains too large. The factorization does not allow to sufficiently reduce the complexity of *Hex* and, in *Blocker Parallel*, the presence of compound actions combined with joint moves for both players brings a large number of combinations.

Note that *LeJoueur* of Jean Noël Vittaut, which won the 2015 Tiltyard Open, is on average 8.5 times faster to ground and factorize the three most complex games (*Breakthrough*, *Hex* and *Blocker Parallel*). This indicates the potential scope for improving these steps.

Table 1 also shows the total number of subgames discovered for each of the 40· games and among them, the ones that are action-dependent and action-independent. The figures in parenthesis indicate the number of discovered subgames considered as useless.

Games at the top of the table are composed of only one action-dependent subgame and sometimes a stepper detected as a useful action-independent subgame. The useless action-independent subgame detected for games like *Breakthrough* or *Sheep and Wolf* corresponds to the *control* fluents which indicate the active player in an alternate moves game and does not represent a playable game per se.

Useless subgames in multiple games are correctly identified. We remark that for *Multiple Tictactoe*, the number of useless subgames is particularly large because these subgames have been over-decomposed as no auxiliary predicate creates a link between their cells.

For the game of *Nim*, our program has detected an action-independent subgame not involved in the end of the game (it is not a stepper) while it is the only subgame useful for the calculation of the score: this is an important clue indicating that this game is impartial.

Except for the special case of Chomp, all the detected subgames are the expected ones and correspond to what would have been obtained by a manual decomposition. Chomp is an example of a game on which the heuristics used for the action effects detection do not work properly. Other actions than eating the poisoned chocolate square have only implicit negative effects which are not detected. These actions are considered as *noop* actions and would be evaluated as equivalent during the game: this could not allow the player to prevent the fatal outcome. Fortunately, such a wrong detection of the action effects is visible in the resulting dependency graph as a huge proportion of fluents and actions are isolated vertices. So we can prevent this error from affecting the game solving.

Table 1. Result of the decomposition for a panel of 40 games descriptions from the servers of Dresden (D), Stanford (S) and Tiltyard (T) with comments on subgames (SG) found.

game	# total of SG	# SG with actions (# useless)	# action-indep. SG (# useless)	time (DNFD)	time (DNF)	comments
Hex (T)	not decomposed after 1 hour					
Blockerparallel (D)	not decomposed after 1 hour					
Asteroids (D)	2	1	1	<1sec	<1sec	
Blocks (D)	2	1	1	<1sec	<1sec	
EightPuzzle (T)	2	1	1	<2sec	<2sec	
Roshambo2 (D)	2	1	1	<1sec	<1sec	
Checkers (D)	3	1	1 (1)	>1hr	<12min	
Breakthrough (T)	2	1	(1)	<16min	<16min	
Sheep and wolf (D)	2	1	(1)	>1hr	<5sec	
Tictactoe (S)	2	1	(1)	≈1sec	<1sec	
Nineboardtictactoe (S)	2	1	(1)	>1hr	<2sec	SG = 9 Tictactoe together
Tictactoex9 (D)	2	1	(1)	>1hr	<5sec	SG = 9 Tictactoe together
Chomp (D)	2	1	(1)	<1sec	<1sec	⚠**Wrong decomposition**
Multiplehamilton (S)	3	1 (1)	1	<1sec	<1sec	SG = Hamilton (only right useful)
Multiplebuttonsandlights (S)	10	1 (9)	1	<1sec	<1sec	SG = group of buttons (only n°5 useful)
Multipletictactoe (S)	75	1 (72)	1 (1)	<10sec	<1sec	SG = Tictactoe n°5 (+ useless SG = cells)
Blockerserial (D)	2	2		<20min	<10min	SG = Blocker ×2
Dualrainbow (S)	2	2		≈1min	<8sec	SG = Rainbow ×2
Asteroidsparallel (D)	3	2	1	<1sec	<1sec	SG = Asteroid ×2
Blocksworldparallel (D)	3	2	1	≈1sec	≈1sec	SG = Blocks ×2
Dualhamilton (S)	3	2	1	<1sec	<1sec	SG = Hamilton ×2
Dualhunter (S)	3	2	1	<2sec	<2sec	SG = Hunter ×2
Incredible (D)	3	2	1	<1sec	<1sec	SG = Maze et Block
Asteroidsserial (D)	4	2	2	<1sec	<1sec	SG = Asteroid ×2
Blocksworldserial (D)	4	2	2	<1sec	<1sec	SG = Blocks ×2
Jointbuttonsandlights (S)	4	3	1	<1sec	<1sec	SG = 3 groups of buttons
LightsOnParallel (T)	5	4	1	<8min	<1sec	SG = 4 groups of lights
LightsOnSimul4 (T)	5	4	1	<8min	<1sec	SG = 4 groups of lights
LightsOnSimultaneous (T)	5	4	1	<8min	<1sec	SG = 4 groups of lights
Nim3 (D)	5	4	1	<2sec	<2sec	SG = 4 heaps + control useful for *goal*
Chinook (S)	6	2	2 (2)	<14sec	<14sec	SG = Checkers ×2
Double tictactoe dengji (D)	3	2	(1)	>1hr	<2sec	SG = Tictactoe ×2
SnakeParallel (T)	3	2	(1)	>1hr	<2sec	SG = Snake ×2
TicTacToeParallel (T)	3	2	(1)	>1hr	≈2sec	SG = Tictactoe ×2
Doubletictactoe (D)	4	2	(2)	>1hr	<1sec	SG = Tictactoe ×2
TicTacHeaven (T)	4	2	(2)	>1hr	<2sec	SG = 9 Tictactoe together + 1 isolated
TicTacToeSerial (T)	4	2	(2)	>1hr	<1sec	SG = Tictactoe ×2
ConnectFourSimultaneous (T)	4	2	(2)	>1hr	<1sec	SG = Connect4 ×2
DualConnect4 (T)	4	2	(2)	>1hr	<1sec	SG = Connect4 ×2
Jointconnectfour (S)	4	2	(2)	>1hr	<1sec	SG = Connect4 ×2

5   Conclusion and Future Work

In this paper we presented a general approach for the decomposition of games described in the *Game Description Language* (GDL). Our program decomposes descriptions of games, compound or not, with any number of players while addressing the problem of joint moves. It decomposes parallel games, games with compound moves and serial games composed of two subgames. It also identifies steppers, useless subgames in multiple games, and unlike previous works, without relying on syntactic elements that can be eliminated by simply rewriting GDL rules. We tested our program on 40 games, compound or not, and have decomposed 32 of them with success in less than 5 s which is a time compatible with GGP competition setups.

Using Meta-action sets is an efficient way to the problem raised by compound moves (Sect. 3.4). However, it requires the completely developed disjunctive normal form of the *next* rules which is computationally expensive. We are seeking another approach to avoid this need or to minimize its computation time. Beside this, we plan to eliminate the ad-hoc heuristics used to identify action effects (Sect. 3.2) and to avoid over-decomposition (Sect. 3.3). We will also address the problem of the decomposition of more than two sequential subgames.

Finally, using these decomposed games to solve the *composition problem* for any games with any number of players remains an open problem.

References

1. Cerexhe, T., Rajaratnam, D., Saffidine, A., Thielscher, M.: A systematic solution to the (de-)composition problem in general game playing. In: Proceedings of the European Conference on Artificial Intelligence (ECAI), pp. 195–200. IOS Press (2014)
2. Cox, E., Schkufza, E., Madsen, R., Genesereth, M.: Factoring general games using propositional automata. In: Proceedings of the IJCAI-09 Workshop on General Game Playing (GIGA 2009), pp. 13–20 (2009)
3. Genesereth, M., Björnsson, Y.: The international general game playing competition. AI Mag. **34**(2), 107–111 (2013)
4. Genesereth, M.R., Love, N., Pell, B.: General game playing: overview of the AAAI competition. AI Mag. **26**(2), 62–72 (2005)
5. Günther, M.: Decomposition of Single Player Games. Master's thesis, TU-Dresden, Germany (2007)
6. Günther, M., Schiffel, S., Thielscher, M.: Factoring general games. In: Proceedings of the IJCAI-09 Workshop on General Game Playing (GIGA 2009), pp. 27–33 (2009)
7. Love, N., Hinrichs, T., Haley, D., Schkufza, E., Genesereth, M.: General Game Playing: Game Description Language Specification. Technical report LG-2006-01, Stanford University, March 2008
8. Schkufza, E., Love, N., Genesereth, M.: Propositional automata and cell automata: representational frameworks for discrete dynamic systems. In: Wobcke, W., Zhang, M. (eds.) AI 2008. LNCS (LNAI), vol. 5360, pp. 56–66. Springer, Heidelberg (2008). doi:10.1007/978-3-540-89378-3_6
9. Vittaut, J., Méhat, J.: Fast instantiation of GGP game descriptions using prolog with tabling. In: ECAI 2014, pp. 1121–1122 (2014)
10. Zhao, D.: Decomposition of Multi-Player Games. Master's thesis, TU-Dresden, Germany (2009)
11. Zhao, D., Schiffel, S., Thielscher, M.: Decomposition of multi-player games. In: Nicholson, A., Li, X. (eds.) AI 2009. LNCS (LNAI), vol. 5866, pp. 475–484. Springer, Heidelberg (2009). doi:10.1007/978-3-642-10439-8_48

Author Index

Printed in the United States
By Bookmasters